SPACE SHUTTLE
A TRIUMPH IN MANUFACTURING

Robert L. Vaughn
Editor

Robert E. King
Manager

Published by

Society of Manufacturing Engineers
Publications Development Department
Marketing Services Division
One SME Drive
P.O. Box 930
Dearborn, Michigan 48121

SPACE SHUTTLE
A TRIUMPH IN MANUFACTURING

SME wishes to acknowledge and express its appreciation to the following contributors for supplying the various articles reprinted within the contents of this book.
Appreciation is also extended to the authors of papers presented at SME conferences or programs as well as to the authors who generously allowed publication of their private work.

American Machinist
McGraw Hill Inc.
1221 Avenue of the Americas
New York, New York 10020

American Metal Market/ Metalworking News
Fairchild Publications
7 East 12 Street
New York, New York 10003

Aviation Week & Space Technology
1777 N. Kent Street
Suite 710
Arlington, Virginia 22209

Compressed Air Magazine
253 E. Washington Avenue
Washington, New Jersey 07882-2495

Richard Harrington
Martin Marietta Aerospace
Michoud Division
P.O. Box 29304
New Orleans, Louisiana 70189

Iron Age
Chilton Company
Chilton Way
Radnor, Pennsylvania 19089

Lockheed Missiles & Space Company, Inc
1111 Lockheed Way
Sunnyvale, California 94086

Machinery and Production Engineering
Findlay Publications Ltd.
Franks Hall, Horton Kirby
Kent DA4 9LL
England

Manufacturing Engineering
Society of Manufacturing Engineers
One SME Drive
P.O. Box 930
Dearborn, Michigan 48121

Materials Engineering
Penton/IPC
1111 Chester Avenue
Cleveland, Ohio 44114

Eugene C. McKannan
George C. Marshall Space
 Flight Center
Materials & Process Laboratory
Huntsville, Alabama 35812

Modern Plastics
McGraw-Hill Inc.
1221 Avenue of the Americas
New York, New York 10020

Research & Development
P.O. Box 1030
Barrington, Illinois 60010

Society for the Advancement of Materials & Process Engineering
P.O. Box 2459
Covina, California 91722

Society of Automotive Engineers
400 Commonwealth Drive
Warrendale, Pennsylvania 15096

Tooling & Production
Huebner Publications, Inc.
6521 Davis Industrial Pkwy.
Solon, Ohio 44138

Welding Design & Fabrication
Penton/IPC
1111 Chester Avenue
Cleveland, Ohio 44114

Welding Journal
American Welding Society
550 N.W. LeJeune Road
Miami, Florida 33125

Cover photo courtesy of Dave Scholl, Lockheed Missiles & Space Company, Inc.

PREFACE

More than two decades of research, planning, advanced design manufacturing and advanced manufacturing technology have gone into the development of the space shuttle of the 80s. The level of sophistication on the space flights has increased so that astronauts no longer wear bulky space suits on-board, and their environment is not a cramped space capsule, but rather a vehicle similar to a commercial airplane. In flight, astronauts breathe air and eat with cutlery. The space shuttle transportation system, itself, is comprised of an orbiter, main engines, external tanks, and a booster made up of two solid rocket motors.

To respond to special requirements and conditions, unique techniques and materials have been introduced to satisfy this growth and development. The articles and papers included in this volume represent some of the most significant and informative literature on space shuttle advanced manufacturing technology. In selecting the pieces for this book, over 400 works on the space shuttle were reviewed. For the most part, the papers included represent state-of-the-art treatments of various aspects of the shuttle make-up. Chapters cover design and assembly, tooling, manufacturing processes, materials, and manufacturing systems.

Even as this book was in its production stages, the space shuttle Discovery set out to launch a communications satellite (by now a routine operation). Discovery's flight made history not for the satellite launch, but rather for the retrieval methods used on two previously mislaunched communications satellites. The dramatic retrieval of the two satellites from useless orbits was accomplished with the orbiter's crane-like robot manipulator arm and the manual maneuvering of the satellites by the astronauts. The satellites were brought back to earth for clean-up and possible reuse. The five-hour process of bringing back satellites into the cargo bay had never been done before. Skills learned from the Mercury, Gemini, Apollo, and shuttle space flights make such procedures a reality.

In addition to the pioneering recovery of the two satellites, every bit of information culled from space missions opens the doors to future experiments and research. Advances in high-level space exploration are encouraged by the United States to benefit both industry and government.

Space shuttle flight tests and evaluations not only result in increased orbit times, retrieval method development, and cargo payload capabilities, but the future of space exploration also lies in the research and experiments that provide day-to-day applications for life on Earth. Research has already begun in the biomedical and health fields on hormone purification, electrophoresis, and cryogenics. Work has been done to perfect optical products and develop crystal growth in space. Experiments will investigate future energy sources and alternatives: how to turn the sun's rays into usable energy, how to reflect light to earth, how to reduce or eliminate crop damage through the use of the sun's rays, how to manufacture superior products in space.

Repercussions of space technology are strongly felt in industry. Space/industrial ties influence industrial procedures from the design and development of new products to the analysis and improvement of manufacturing processes, parts, and materials for higher productivity; stronger quality control; and better management and software systems.

Products evolving from the space program result from direct use in space or through the technology developed from the program. Sometimes the product already existed, but wasn't commonly used until it encountered a space-related problem. An example is metallized materials. Designed for use in space, the vacuum-vaporized metallized material has now been adapted to applications on Earth. Manufacturers use the metallized material in producing their own products and also sell rolls of the material to other manufacturers.

The industry of liquid hydrogen production was greatly expanded as a result of research and tests ordered by the Air Force for use in space. Intended as a fuel for rockets and missiles, the resultant information has reduced production costs of liquid hydrogen and has made large quantities available for commercial use.

There are several conditions that exist in space that do not exist on Earth, namely: a cheap supply of solar energy, a near-perfect vacuum, extreme temperatures, and an environment free from atmospheric-related corrosives and impurities. The condition in space of greatest value is the almost total absence of gravitational force.

These qualities offer the chance to test processes that cannot be tried on Earth. Materials processing operations could benefit from the microgravity of space to produce superior metal alloys, pure glass, flawless crystals, and high-purity pharmaceuticals. By eliminating or reducing the adverse effects of gravity, space processing could encourage entirely new industries.

Space exploration is not simply the quest for knowledge, but it is also the ability to turn the unknown into a reality that will improve and protect the future of mankind.

I would like to thank all of the companies, organizations, publishers, and authors who gave permission to have their articles reprinted in this volume. Thanks also to the Publications Development Department staff at SME for their assistance in the research and development required in making this book possible.

Robert L. Vaughn, CMfgE, P.E.
Lockheed Missiles & Space Company Inc.

ABOUT
THE EDITOR

Robert L. Vaughn is Director of Productivity for the Lockheed Missiles & Space Company, Inc. He holds a BS in Mechanical Engineering from Iowa State University and an MS degree in Administrative Management from California State University at Northridge. He is a Certified Manufacturing Engineer and a Registered Professional Engineer in California.

Having joined the Lockheed Corporation in 1953, Mr. Vaughn has held responsibilities in materials and process engineering, methods engineering, production design, research and development, and design assurance. In

1977, he was named Chief Manufacturing Engineer of Lockheed's Space Systems Division and was promoted to Director of Productivity for Lockheed Missiles and Space Company in 1981.

Mr. Vaughn is internationally recognized in the field of metal removal, particularly that of machining and research in the high temperature, high strength alloys required in the aerospace industry. He holds five US patents in the machining field. He is author or co-author of more than 70 published technical articles or books and has lectured extensively in North America, Asia and Europe.

For these technical contributions he received several awards including the Society of Manufacturing Engineer's (SME) 1969 Gold Medal and awards from the California Society of Professional Engineers, the San Fernando Valley Engineers Council and the Society of Automotive Engineers, Aerospace Division. He has received special commendations from the California State Legislature and from the California State Polytechnic University.

As a member of SME's Board of Directors from 1972 to 1983, Mr. Vaughn has also served as International Secretary, Treasurer, Vice President, and President (1981-1982). He has been highly active in SME technical and administrative committees on the national, regional, and chapter levels. He was Chairman of SME's National Engineering Conferences, Programs, Engineering Materials, and Manufacturing/Design Assurance Committees.

In addition to membership in SME, Vaughn is a member of the American Society for Metals, the National Management Association and the National Society of Professional Engineers. He is a Fellow of the American Society of Mechanical Engineers, the Institute for the Advancement of Engineering and the Institution of Production Engineers. He is also a member and Past President of the San Fernando Valley Engineers Council and a Past President and Director of the San Fernando Valley Chapter, California Society of Professional Engineers.

SME

The informative volumes of the Manufacturing Update Series are part of the Society of Manufacturing Engineers' many faceted effort to provide the latest information and developments in engineering.

Technology is constantly evolving. To be successful, today's engineers must keep pace with the torrent of information that appears each day. To meet this need, SME provides, in addition to the Manufacturing Update Series, many opportunities in continuing education for its members.

These opportunities include:

- Monthly meetings through three associations and their more than 270 chapters which provide a forum for member participation and involvement.

- Educational programs including seminars, clinics, programmed learning courses, as well as videotapes and films.

- Conferences and expositions which enable engineers and managers to examine the latest manufacturing concepts and technology.

- Publications including the periodicals *Manufacturing Engineering, Robotics Today,* and *CAD/CAM Technology,* the *SME Newsletter,* the *Technical Digest, Journal of Manufacturing Systems,* and a wide variety of text and reference books covering everything from the basics to manufacturing trends.

- The SME Manufacturing Engineering Certification Institute formally recognizes manufacturing engineers and technologists for their technical expertise and knowledge acquired through experience and education.

- The Manufacturing Engineering Education Foundation was created by SME to improve productivity through education. The foundation provides financial support for equipment development, laboratory instruction, fellowships, library expansion, and research.

- A database, accessible through SME containing Technical Papers and publication articles in abstracted form. The information on Technology In Manufacturing Engineering database is only one of several accessible through the Society.

SME is an international organization with more than 71,000 members in 65 countries worldwide. The Society is a forum for engineers and managers to share ideas, information, and accomplishments.

The Society works continuously with organizations such as the American National Standards Institute, the International Organization for Standardization, and others, to establish and maintain the highest professional standards.

As a leader among professional societies, SME assesses industry trends, then interprets and disseminates the information. SME members have discovered that their membership broadens their knowledge and experience throughout their careers. The Society is truly industry's partner in productivity.

MANUFACTURING UPDATE SERIES

Published by the Society of Manufacturing Engineers, the Manufacturing Update Series provides significant, up-to-date information on a variety of topics relating to the manufacturing industry. This series is intended for engineers working in the field, technical and research libraries, and also as reference material for educational institutions.

The information contained in this volume doesn't stop at merely providing the basic data to solve practical shop problems. It also can provide the fundamental concepts for engineers who are reviewing a subject for the first time to discover the state-of-the-art before undertaking new research or application. Each volume of this series is a gathering of journal articles, technical papers and reports that have been reprinted with expressed permission from the various authors, publishers or companies identified within the book. SME technical committees, educators, engineers, and managers working within industry, are responsible for the selection of material in this series.

We sincerely hope that the information collected in this publication will be of value to you and your company. If you feel there is a shortage of technical information on a specific manufacturing area, please let us know. Send your thoughts to the Manager of Educational Resources, Marketing Services Department at SME. Your request will be considered for possible publication by SME—the leader in disseminating and publishing technical information for the engineer.

TABLE OF CONTENTS

CHAPTERS

3 DESIGN AND ASSEMBLY

4 TOOLING

5 MANUFACTURING SYSTEMS

CHAPTER 1

THE SPACE SHUTTLE TAKES OFF

From artificial organs and alternative energy systems, to new products and processes, reapplied space technology helps improve the quality of life on Earth.

Space Technology Comes Down To Earth

For many, the concepts of moon-walks, manned satellite linkups, close-up photography of distant planets and other space projects seem far removed from daily life. In reality, the technology that accompanies the space program does relate to Earth problems when it is reapplied to serve individuals and industry.

Through its policy of technology transfer, the National Aeronautics and Space Administration makes available to the private sector the vast storehouse of knowledge that accumulated during its 25-year existence. According to NASA, "The program aims to broaden and accelerate technology transfer to create new products, new processes and new jobs, and thus gain substantial dividends on the funds invested in aerospace research." Areas in which this space-gained knowledge exist are widely diverse.

A highly identifiable connection between Earth and space in daily living is the use of communications satellites. This space hardware—satcoms—was pioneered by NASA over 20 years ago, establishing a technology base (much of which is too high risk for private investment) that paved the way for commercial use of satellites. Since the 1960s, the satellite population boomed, and now NASA works on technology that would make satcoms more efficient to avoid a "traffic jam" in space.

NASA assumes the responsibility for satellite technology because these spacecraft provide a wide range of services in the public interest. More than just improving TV programming, satcoms, such as the Landsat Resources Survey Satellite, offer tremendous aid in Earth management. Landsat monitors changing conditions on the earth's surface by a process known as remote sensing, in which spaceborne sensors detect various types of radiation emitted or reflected from objects on earth. This information can apply to mineral and petroleum exploration, agricultural crop forecasting, rangeland and forest management, mapping, land use management, water quality evaluation, and disaster assessment.

NASA is presently investigating the possibilities of space platforms and stations that would perform the functions of several satcoms. One study under research with the Department of Energy (DOE) is the potential Satellite Power System. This involves miles-long space platforms containing devices to transform sunlight into electricity, and

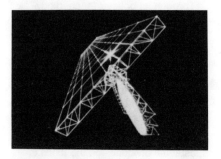

Artist's rendering depicts solar power satellite under study by NASA and DOE. It would provide 5000 MW of electrical power to a ground receiving station for introduction into a power grid.

then microwave the energy to earth receivers. One space platform could supply the power for a large city; a complete system could relieve many of the nation's power problems.

NASA reports that these platforms and space stations could be used industrially also, creating the opportunity for in-space manufacturing. Several materials processing operations would

benefit from the microgravity of space to produce superior metal alloys, pure glass, flawless crystals, and high-purity pharmaceuticals. By eliminating or reducing the adverse effects of gravity, such as the reluctance for certain metals to mix, inducing impurities, and limiting output in pharmaceutical preparation, space processing could open the door to whole new industries.

Closer to home

Aside from space flight, NASA is involved with technology that addresses problems of the aviation industry. Designed to cope with challenges facing the commercial airline industry, the Aircraft Energy Efficiency Program (ACEE) studies ways to reduce jet fuel consumption. It looks at elements that affect fuel efficiency, such as engine design, aerodynamics, structures, guidance, and control. ACEE research projects have brought modest to dramatic fuel cost savings. It is estimated that this new technology could cut fuel consumption as much as 50 percent. The effort to increase the efficiency of jet aircraft is well motivated, since a reduction of just 10 percent per annum could save more than $1 billion for the commercial airlines industry alone.

Fuel efficiency is of equal concern in ground propulsion research sponsored by NASA. The bulk of this research centers on reducing fuel consumption, combined with efforts to wean the auto from single fuel dependency.

Experimental projects show that several viable designs offer increased efficiency and fuel alternatives. One such project is the Stirling propulsion system developed by United Stirling of Sweden for Mechanical Technology Inc. This research deals with an automotive engine that burns a conventional air/fuel mixture, but features a continuous com-

Stirling engine, featuring a continuous combustion heater head located outside the cylinders, lowers fuel consumption and can operate on a variety of fuels.

bustion heater head located outside the cylinders. The design promises lower fuel consumption—from 30 to 50 percent—and can operate on a variety of fuels: gasoline, kerosene, alcohol, diesel fuel, butane and others, including future synthetics.

Managed by Lewis Research Center, this program also investigates developing automotive gas turbine engines. Cars thus equipped would be more efficient, less polluting, and could accommodate a variety of fuels.

Other engine research programs are underway. Beech Aircraft Corpora-

Liquid methane tank, based on spacecraft technology, allows cars and trucks to convert to methane fuel.

Space Technology

tion's Boulder Division, borrowing from aerospace fuel technology, has developed a system that converts standard cars and trucks to accept methane fuel, from a liquid methane tank, rather than gasoline. Cleaner and half as costly as gas, liquid methane is a "cryogenic" fuel requiring a storage temperature of −260° F. Technology made available from the liquefied gas storage systems developed for the Apollo and Skylab spacecraft and the Space Shuttle led the way for the liquid methane tank used in the Beech conversions.

Energy savings and generation

Although the Satellite Power System is a future consideration, a great number of energy-producing devices and energy-saving technologies and products already exist that derive directly from space research and development.

Probably the most "pure" example of space-derived energy source development is the photovoltaic cell. It emerged as the most efficient and practical power source to operate equipment aboard spacecraft since it is self-contained and light weight. Fueled by sunlight, no "conventional" fuel need be carried. Solar cells have no moving parts, emit no pollution, and consist of one of the most abundant substances on earth—silicon.

Although applicable to many earth uses, solar cells presently are too costly for use except where absolutely necessary. Early in the space program, the cost per watt of electric power was hundreds of dollars. This has been reduced to under $10 per watt for current use, and DOE hopes to reduce the cost per watt to below $1, making it feasible for private and commercial energy source use.

NASA is investigating parabolic dish concentrators at its Parabolic Dish Site. These concentrators collect and focus sunlight to generate electrical power, or they can generate high temperatures for industrial processes. When perfected, individual units could serve small communities, factories or farms, or groups of the collectors could produce as much as a million watts of electrical energy for large industrial and utility applications.

Other energy research investigates wind turbines with capacities to 4 million watts, and coal-fired power plants that would turn high-sulphur coal into clean-burning gas.

Aluminized "heat shield" insulation served as a radiation barrier in the Apollo spacecraft.

Working hand-in-hand with solar energy research is the space-derived technology for improving insulation. Applying techniques learned in the Apollo Program, which coped with temperature shifts of up to 800° F, the Watt Count System, developed by Arnold Engineering Development Center, brings this energy efficient know-how into the home. The system utilizes aluminum "heat shield" insulation, which erected a radiation barrier on the spacecraft, to provide as much as 45 percent energy savings in homes.

Industrial applications

Industry probably feels the strongest and most widespread impact of space technology spinoff. The space-industrial connection pervades every phase of the industrial process from developing new products and improved designs, to manufacturing processes, parts, and materials for increased productivity, industrial power systems, improved quality control, building design and improved industrial environments, and computer software and management systems.

Products emerging from the space program either trace directly from space use, or the technology the program developed. On occasion, the product pre-existed, but had not been widely employed until it was applied to a space-related problem. An example is metallized material.

During the design stages of Echo 1, the first NASA communications satellite, metallized mylar polyester coated with aluminum made it possible to build a reflective balloon that at launch folded to "beach ball" size, and in orbit inflated as tall as a ten-story building. The attendant research led to this material's use on nearly all U.S. spacecraft, primarily as a thermal insulator. Advances in the technology of vacuum-vaporized metallization resulted in a reborn industry with great growth potential. It originally saw only limited duty as a decorative item.

King-Seeley Thermos Company, (KST) Metallized Products Division, produces a large product line stemming from the NASA-sponsored research of the late 50s. Building on that technology base, KST developed new and advanced processes for metallization, resulting in coatings of gold, silver, copper and zinc, as well as aluminum. KST uses these materials in its products, and also supplies them in rolled sheet form to other manufacturers. Today, this metallized material appears in such products as outdoor apparel, food packaging, insulating wall coverings, aircraft covers, and photographic reflectors. It also forms the basis for the "Space" blanket, a light, compact blanket that retains body heat, and is carried by paramedics, hikers, athletes, and others for emergencies. These reflective blankets, along with freeze-dried foods, are among the highest profile space-spurred products.

Similarly, the needs of the space program expanded another industry—the production of liquid hydrogen—by prompting research. Liquid hydrogen, like metallized materials, existed in limited supply in the 1950s, when the U.S. Air Force began investigating its use as a fuel for missiles and rockets. Prior to the space program's demands, the USAF required the liquefied gas in volumes hundreds of times greater than the technology of the day could produce. Assigned to develop high quantity production of liquid hydrogen was Air Products and Chemicals, Inc.

By the end of the 1950s, three Air Products plants supplied liquid hydrogen to the USAF, the largest producing 30 tons a day. Through the next two decades, the company ultimately opened three more plants, each with a 30-ton-a-day capacity, to meet space program needs. During this time, Air Products improved the process for high-volume production, developed storage and handling technology, and lowered production costs. This last element opened the door to increased commercial use of liquid hydrogen.

With increased availability and the means for transporting and storing this cryogenic material—it must be maintained at $-400°$ F—liquid hydrogen entered into the processes of petroleum refineries, chemical and pharmaceutical companies, food processors, metal manufacturers, electronics companies, and electric utilities. Recently, Air Products opened a facility solely to supply commercial users. From the technology base acquired in responding to the needs of space, a new product emerged that improved the productivity, efficiency, and quality level of many industries.

Other examples of cryogenic technology transfer involve the storage of liquefied natural gas (LNG) and liquid methane, both requiring $-260°$ F maintenance temperatures. Chicago Bridge and Iron Company incorporated NASA research information, which addressed cooldown of cryogenic transfer lines, in its design for ship-to-shore transfer and storage systems of LNG. Methane storage technology developed for NASA's Apollo and Skylab spacecraft allowed experiments in methane-powered automobiles, as previously discussed.

The transfer of testing and analytical technologies proves to be one of the most beneficial to industry. For example, as spacecraft incorporated more composite materials, Lewis Research Center developed a means of testing for strength characteristics. Combining ultrasonics with new acoustic emission technology, the method can accurately predict strength analysis and response to stresses. This testing procedure will benefit those industries investigating composite material use, such as automotive and machinery manufacturers.

PIND (Particle Impact Noise Detection) devices, which test for conductive particles that could cause system failures in integrated circuits, are being produced for use by semiconductor manufacturers, users, and test labs. Pressure Systems Inc., has designed a digital pressure transmitter that increases productivity in industrial processes requiring quick, accurate multiple pressure measurements. Both of these devices take advantage of NASA research and technical information.

Important to the mining and agricultural industries, among others, is the Hand Held Ratioing Radiometer, a sensory system for measuring and analyzing reflective radiation. Developed by Jet Propulsion Laboratory, and now a commercial product of Barringer Research Ltd., this device operates in the field and can determine the specific mineral presence in rock, the health status and potential yield of crops, and other sensory information.

Also of benefit to the mining industry and to any situation where bolts come under severe stresses, is a NASA-developed ultrasonic bolt stress monitor that Langley Research Center produced for industrial use. The device, called Pulse Phase Locked Loop (P^2L^2), particularly adapts to testing mine roof bolts.

Space-derived technology not only aids in testing industrial products and processes, but helps improve many methods of industrial production. An example is laser balancing of rotors found in turbine machinery. Developed jointly by Lewis Research Center and Mechanical Technology Inc., this method replaces the manual procedure of spin testing, adding or subtracting weight, and rechecking. The laser offers precision, speed, and decreased costs to the manufacturer.

Software

NASA offers to almost every American business the opportunity to share in the benefits of space technology through its extensive library of computer programs.

A large number of industries, regardless of end products—from children's toys to giant turbines—share common problems and costs involved in design research and manufacturing of their end products. One approach to overcoming these problems and increasing efficiency is to computerize. Recognizing that the software involved with computer operations can represent a major expense, NASA maintains a bank of over 1500 programs available to private industry. These programs, originally developed by NASA, the Department of Defense, and other government agencies, are available through the Computer Software Management and Information Center (COSMIC) located at the University of Georgia. The programs reapply to industrial situations, often solving problems far removed from their original task.

Ingersoll-Rand Company is one of the firms that avails itself of these programs. The company's Pump Group employs a program titled MERIDL, which performs flow analysis calculations. With the assistance of the I-R Research Center, Princeton, N.J., the program helps design impellers for large circulating pumps. The company uses an additional program called TSONIC for analyzing flow velocities in pumps, compressors, and turbines.

Another COSMIC program called NASTRAN (NASA Structural Analysis) is used widely in design engineering. This general purpose program analyzes a design and projects how it will function under all possible situations. Used by many companies, the program has assisted Magnavox Government and Industrial Electronics Company in vibration testing of electronic components, as well as other testing procedures; it has helped Tennessee Eastman analyze equipment failure to avoid costly downtime; Ford Motor Company has analyzed new auto designs with it to reduce prototype costs.

Besides direct involvement in the operations of American industry, NASA technology helps improve the industrial environment. Benefits include noise abatement research and technology, efficient and economical night lighting,

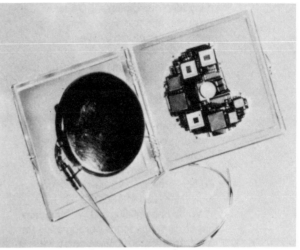

About the size of a woman's compact, Programmable Implantable Medication System continuously delivers medication from a source within a patient's body.

In designing the impeller for the 73APHJ large vertical circulating pump, Ingersoll-Rand made use of a NASA-developed computer program that performs flow analysis calculations.

and products for more attractive and productive office design. In fact, the management system of aerospace operations, which must efficiently incorporate many diverse elements, is being studied and applied to complex industrial operations.

Good health

One of the most important areas of space technology transfer lies in developing human health aids. Many dramatic health devices have come about because space program researchers made advances in "microminiaturization." In turn, devices applicable to an environment even more critically demanding than that of space—the human body—have been developed.

The Programmable Implantable Medication System (PIMS) is a brilliant example of such technology. This system is a cooperative effort of the Applied Physics Laboratory (APL) of Johns Hopkins University, the Goddard Space Flight Center, and several commercial manufacturers. As described in a NASA annual report, "PIMS is a microminiaturized, computer-directed system for continuous delivery of medication to target organs, in precisely con-

trolled amounts, from a source within the patient's body."

Featuring an Implantable Programmable Infusion Pump, the PIMS offers patients, most notably those requiring treatment for diabetes, a finely controlled and monitored administration of medication—a method considered superior to injection. Once implanted, the infusion pump, together with its minicomputer for dosage control, medication reservoir, internal tubing for medication delivery, and lithium battery, is controlled from outside according to the patient's needs.

In coronary treatment, the Pacer—a rechargeable cardiac pacemaker developed by PAL and produced by Pacesetter Systems, Inc.—communicates by telemetry to the physician the status of the patient's heart, and allows reprogramming accordingly. Telemetry, used also in the PIMS, is the wireless process by which ground control queries orbiting spacecraft. Coded demands and response data convert to electrical signals that a receiver presents as informational reports.

Representing a major breakthrough in heart-assist devices, this cardiac pacing system eliminates the need for

recurring surgery for replacing batteries after initial implantation. The Pacer's rechargeability directly derives from the technology surrounding spacecraft electrical power systems.

Commercial products and processes

Not all space-derived technology applies to profound end products. Many comfort and leisure items also trace from space know-how. For example, everyday cordless appliances and tools stem from technology gained when a cordless lunar drill was developed for Apollo astronauts to take core samples from the moon's surface.

Heat pipe technology, developed by Los Alamos Scientific Laboratory, aided James M. Stewart's experiments that lead to his patented "head tubes" for improved plastics production. With a license from Stewart, Kona Corporation developed the Kona Nozzle, which improves efficiency and safety of plastics manufacturing equipment. From original technology aimed at solving the problem of distributing heat in a nonrotating satellite, emerged improved production techniques for a wide range of plastic items. Included are pens and lighters from Bic Pen Corporation, parts for Kodak and Polaroid cameras, Tupperware kitchenware, Ford Motor Company auto components, RCA television cabinets, and telephones and components manufactured by Western Electric.

Recreational items borrow from the space program too. In the air, hang gliders made by U. S. Moyes, Inc., employ certain NASA technology that traces from the Administration's predecessor group, the National Advisory Committee for Aeronautics. On the sea, yacht racers are aided by a French-American space-based monitoring system called ARGOS.

While hang gliding and yachting may not be common forms of recreation for most Americans, donning sportswear is certainly widespread. One company, Techni-Clothes Inc., borrows from space technology to drape the sports participant's body. Its founder, Dr. Lawrence H. Kuznetz, a bioengineer and physiologist at Johnson Space Center, applied his knowledge of liquid cooling systems for space suits in designing a line of sports apparel that transfers heat away from the wearer's body through a heat-absorbing gel. The product fits well in events such as long-distance running and jogging.

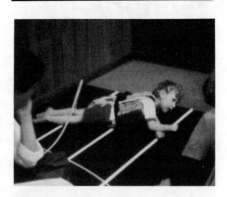

Devices to simulate weightlessness find use in medicine and therapy, as shown here helping a handicapped youngster learn to crawl.

It is clear that in its reapplied state, space technology contributes significantly to improved quality of life on Earth. □

SPACE SHUTTLE TILE— THE EARLY LOCKHEED YEARS

Wilson B. Schramm
Product Development Representative,
Advanced Fiber Ceramic Composite Materials and Applications

Ronald P. Banas
Program Engineering Manager, Thermal Protection Systems

Y. Douglas Izu
Senior Staff Engineer

Lockheed Missiles & Space Company
Space Systems Division

COLUMBIA RETURNS

The time was 11:50am CST on 14 April 1981—Mission Commander John W. Young and Captain Robert L. Crippen piloting *Columbia* descend through Entry Interface at 400,000 feet and Mach 24 over the Pacific somewhere east of Guam. Now committed, they can feel the first nibble of Earth's atmosphere dragging them down to a landing at Edwards Air Force Base, 4390 miles and only a few minutes away. Ahead is the first searing reentry and the first functional test of Lockheed's ceramic tile heat shield.

Twenty years had passed since Commander John Glenn and the first U.S. manned orbiting Mercury spacecraft splashed down in the Atlantic Ocean, but that craft and all subsequent Mercury, Gemini, and Apollo spacecraft had employed ablative shields to counter the heat of reentry into the earth's atmosphere. Thus, Lockheed's ceramic tile—an entirely different concept—must have been on their minds as Young and Crippen waited for aerodynamic control forces to build up—and for temperatures to climb above 2000°F in the thin glass and

ceramic skin beneath their feet. This flight represented a lot of "firsts," but perhaps the principal one was the thermal protection system that took 20 years to develop, and was now bleeding off *Columbia*'s kinetic energy and reradiating the intense heat back to deep space and the earth.

The words that the world was waiting for came minutes later as the spacecraft emerged from radio blackout: "Hello Houston, *Columbia* here, we're doing Mach 10.3 at 188." It worked. After a flawless touchdown on the dry lake bed at 12:22pm CST, John Young emerged and ducked away from the cameras for a quick look at the forward underbody heat shield. Young later spoke of his feelings: "We had done it! The whole package. There was not one tile missing. Not one! Considering how many human beings worked on that rascal, the complexity of putting those tiles on, and the beating they took, that's a wonder. It's the finest bricklaying job that's ever been done."

Lockheed provided the bricks for what became affectionately known as the "Flying Brickyard." NASA's fleet of Space Shuttle orbiters, *Columbia, Challenger, Discovery,* and *Atlantis.* In

a later visit to the plant at Lockheed Missiles & Space Company (LMSC) in Sunnyvale, John Young said, "The tiles were a remarkable achievement. It's just hard to believe that something like that worked as well as it did. Absolutely remarkable."

The tiles were there because their unique physical properties and thermal shock tolerance provided the only heat shield concept with a positive payload margin for the Space Transportation System (STS). Their performance has been exciting!

When the program is completed LMSC will have manufactured 100,000 glass-coated fiber ceramic tiles with the precision required to form the aerodynamic surface configuration of the Space Shuttle orbiter. These lightweight insulating tiles meet the exacting requirements for Space Shuttle missions, routinely returning this $1-billion spacecraft to earth through the searing heat of reentry at peak temperatures well above 2000°F. This program spans from the early research and development in 1962 through large-scale production and product improvement—representing 20 years of continuous technical, management, and manufacturing experience with advanced engineering materials and processes.

LMSC delivered a total of 22,158 shipset tiles plus all materials for the remaining 8599 closeout and special tiles for each installation on *Columbia* and *Challenger*, the first two orbiters flying in NASA's Space Shuttle fleet. Each tile in *Columbia*'s 30,757 shipset has a distinct individual configuration with a specific part number, machined with diamond tools on numerical-controlled mills from a computer-generated master dimension data system and part program.

There is a lot more to the Thermal Protection System (TPS) than tiles, which cover about 70 percent of the orbiter surface. Other key elements of the system include the quartz windshields, the Reinforced Carbon-Carbon (RCC) leading edges and nose cap, flexible hinge-line thermal barriers, and the carrier plate attachment systems. Rockwell International Corporation as prime contractor has the responsibility for these elements as well as responsibility for thermal and structural design of the tile and its attachment system.

The first shipset of rigidized silica-fiber tiles installed by Rockwell International, was manufactured with two

Immediately after Columbia was "safed" following its first flight, Mission Commander John Young (left) did a fast walkaround to check the condition of the thermal protection system tiles.

Columbia hangs vertically in the Vehicle Assembly Building at Kennedy Space Center.

Crew of STS-2 inspect the thermal protection system tiles on Shuttle nose after landing.

product densities, LI-900 at 9-lb/cu ft density and LI-2200 at 22-lb/cu ft density. The tiles are provided with two borosilicate glass coatings: the familiar black High-temperature Reusable Surface Insulation (HRSI), which predominantly covers the underbody, and the white Low-temperature Reusable Surface Insulation (LRSI), which fills in the upper surfaces. The unique characteristics of this successful reradiative thermal protection system stem from unsurpassed thermal-shock resistance and light weight achieved with an amorphous silica-fiber composite structure, and the compatible high-emittance borosilicate glass coating systems. Delivery of a third shipset for *Discovery* has been completed and manufacturing effort for the final *Atlantis* shipset is well ahead of schedule.

PROGRAM HIGHLIGHTS

Fail-Safe Performance The widely held perception that tiles are fragile and break like a sheet of glass is simply not correct. Properly engineered and properly handled the tiles are remarkably tough. There has not been a single thermally induced tile failure in flight. There has been damage from external tank ice impact at launch, runway debris, hailstorms, frozen moisture, and manhandling. Tile separation experienced early in the program was associated with the attachment system. Ice impact damaged tiles have shown remarkable fail-safe reentry characteristics.

Turn-Around Early experience with *Columbia* showed tile removals due to in-flight damage dropped dramatically from 250 after STS-1 to 120 after STS-5. The first flight of *Challenger* required replacement of 11 damaged tiles, mostly from ice impact damage. A replacement tile can be manufactured in as little as two days from receipt of identification, and literally "dialed up" on the computer-driven numerically controlled mills from a master dimension data base in residence at LMSC in Sunnyvale.

Why Tiles? The ceramic insulation system must be subdivided into individual tiles to accommodate the complex contours of the aluminum airframe as well as to absorb large differential thermal expansions over extremes of temperature during the mission—from minus 170°F in extended orbital cold-soak conditions to 2300°F during reentry. Pure silica LI-900 tile spans are typically in the 6-to-10-inch range. The Strain Isolation Pad (SIP) and bond system supports the tile in flight, absorbs the internal cold soak thermal differential strain, and floats the tile on the aluminum airframe structure as it flexes under dynamic loading conditions.

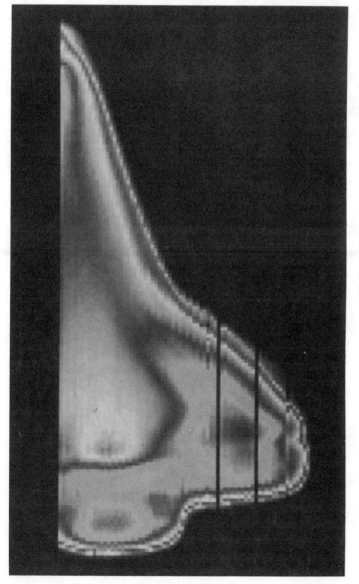

This image of the Space Shuttle Columbia's heating patterns was taken with an airborne infrared telescope as the orbiter reentered the earth's atmosphere on 30 March 1982 after its third flight into space. Mounted on a NASA/Lockheed C-141 StarLifter aircraft, the telescope caught Columbia during reentry when it had a speed of Mach 15.6 and an altitude of 185,300 ft.

Tiles installed on nose cap illustrate contouring requirements of tile manufacture.

View of the underside of Columbia nearing the successful completion of STS-2 during the approach for a landing at Edwards Air Force Base in California.

How It Works The tiles accomplish three major tasks:

1. The tiles dissipate 90 percent of the reentry heat energy in reradiation back to the earth's atmosphere and deep space from the Reaction Cured Glass (RCG) coating which is only 12 mils thick. The energy is transformed into surface heating at incandescent temperatures, and is dissipated because of the high emissivity inherent in the coating's optical properties. The RCG coating forms the aerodynamic skin of the Orbiter vehicle and rapidly heats to 2000°F for ten minutes during reentry. The peak heating rate typically is 25 Btu/sq ft during this critical period.

2. The tiles provide the unique fused silica fiber insulation structure that allows a rapid surface temperature response and supports the glass aerodynamic skin of the vehicle. The structural insulation delays in-depth temperature response and takes the thermal shock and acoustic impulses. Each fused silica tile transmits the aerodynamic loads into the orbiter airframe structure through a Strain Isolation Pad. About 5 percent of the incident energy is dissipated by convection back to the cooler atmosphere during let-down for a landing.

3. The tiles delay the remaining 5 percent incident energy from reaching the aluminum orbiter airframe structure until after the landing is completed. The low thermal diffusivity, a unique product of density, specific heat, and thermal conductivity delays this heat pulse typically for 30 minutes while the aluminum structure never exceeds 350°F.

BEGINNINGS

The Space Shuttle TPS evolved over many years and has resulted from the contributions and dedication of many individuals and organizations, including Rockwell International Corporation and the NASA research centers, as well as Lockheed Missiles & Space Co., Inc.

Radomes During the early 1960s, LMSC conducted research and development of specially designed and integrated ceramic composite structures based on inorganically bonded ceramic fibers. These structures included compatible inorganic binders for both continuous-filament-wound fibers and cast staple fibers. Centrifugal slurry-casting techniques were demonstrated as well as conventional slurry-casting tower processes for the fibrous class of materials. New concepts for TPS structures evolved for antenna and radome applications, combining structural load-bearing functions and electromagnetic transmission functions. The concept of a reusable rigid fiber ceramic insulation material as a reentry vehicle heat shield was first identified in an LMSC patent disclosure in December 1960.

As early as 1962 a conical Apollo radome was built of fused silica composites to withstand the reentry environment. The external shell of this radome was filament-wound with pure

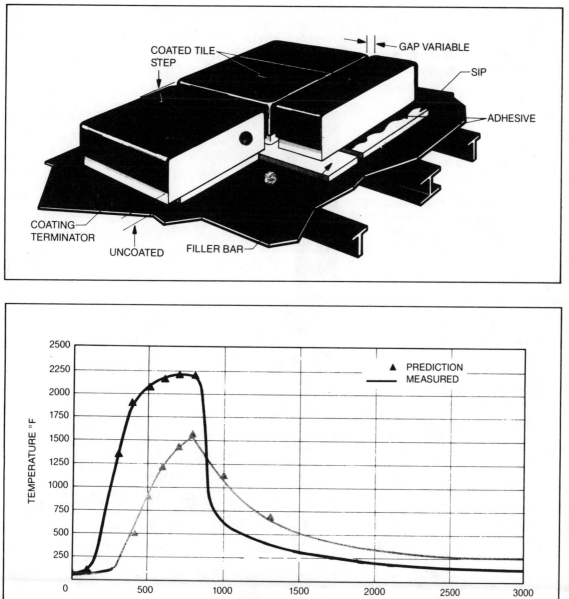

COATED TILE
STEP

GAP VARIABLE

SIP

ADHESIVE

COATING
TERMINATOR

UNCOATED

FILLER BAR

*Representative thermal
protection system cross
section showing tiles
bonded to an aluminum
airframe structure.
Colored dots indicate
locations that relate to
temperature/time curves
in illustration below.*

▲ PREDICTION
— MEASURED

TEMPERATURE °F

TIME IN SECONDS

*Predicted temperature
responses correspond
closely with NASA-JSC
600-second reentry
simulation tests of LI-900
tile #007. See above
illustration for
temperature-response
locations.*

silica fiber. A lightweight internal insulation was centrifugally cast from short staple silica fiber, and the two sections densified with colloidal silica and sintered into a composite structure. This early fiber ceramic reinforced composite structure was 100-percent inorganic and non-ablative. The design requirements changed during evolution of Apollo, however, and it never flew.

Impregnation During the early 1960s ballistic and lifting entry vehicles under consideration were typically small in scale and had sharp nose cap and leading edge radius geometry. Characteristically, peak heating rates were very high

and the only suitable materials at the time were either heat sink or ablators. Initial attempts at inorganic ceramic composites designed specifically for ballistic and lifting entry thermal protection were limited to high-density solid refractories such as zirconia used in Dynasoar.

An early LMSC development designated Lockheat was based on the creation of a low-density inorganically bonded refractory fiber skeleton structure having controlled porosity and microstructure. These structures were impregnated with non-charring organic coolants to achieve a transpiration cooling function at extreme entry heat

flux rates. A large matrix of alternative fiber and binder materials was explored in the early years, including metallic and inorganic metal-oxide fibers and a variety of metal-oxide binders. Silica, alumina, and boria remain the principal components carried over into today's materials. Multi-layered composites of pure silica filament-wound and staple-fiber materials were produced with coolant impregnants tailored to meet various specific weights in the range of 2.5 to 5.0 lb/sq ft. Very acceptable backface temperature response was demonstrated during the early 1960s in reentry simulation tests of typical composite heat shield designs.

Antecedents of Space Shuttle These early investigations of silica fiber ceramics that evolved into the Space Shuttle TPS were well under way at the time of John Glenn's historic first U.S. manned space flight in Mercury—inspired by a clear insight into the potential for a routine multiple reuse lifting entry spacecraft in the future. In 1965 a specific formulation of silica-bonded staple fibers was selected for development to meet the requirements for lifting entry applications. This silica formulation was LI-1500 at a density of 15 lb/cu ft. Initially, the 89-percent porous silica composite was impregnated on the external surface to a pre-determined depth with an organic impregnant for protection against heating above the "melting point" of silica. "Cooling" resulted at high heating rates because of the endothermic decomposition and transpiration of the selected organic constituent. After the organic component was depleted, the remaining silica-fiber insulation provided continued protection for the duration of the more moderate heating phases.

TRANSITION

The late 1960s marked an intensive period of advancement in lifting entry state of the art as well as preliminary design and system definition among the many government research centers and their aerospace contractors. The Lockheed-California Company advanced design organizations became seriously involved in lifting entry aerodynamic configuration development and preliminary design of upper-stage vehicle applications. At LMSC, advanced design teams were following different approaches, utilizing cryogenic ascent propulsion systems integrated with lifting entry aerodynamic configurations to achieve characteristics of a true Space Transportation System.

These advanced concepts all needed a "magic" heat shield—light weight, tough, reusable. Up to this time, heat shields typically consisted of super alloy metallic skins, reinforced honeycomb ablators like Mercury's heat shield, or impregnated ablative silica-phenolic fiber and fabric composites. These con-

cepts were backed up with fabric and foil-covered fiber insulation blankets and panels to insulate underlying vehicle primary structure and subsystems. Weight was a killer.

The connection was made at LMSC in February 1966 over a block of LI-1500 fused silica material on the conference room table. The primary heat shield concept turned around completely with

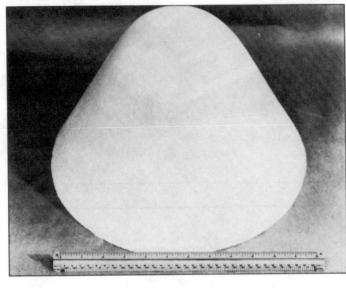

This conical Apollo radome was built of fused silica composites in the early 1960s. This early ceramic structure was 100 percent inorganic and non-ablative.

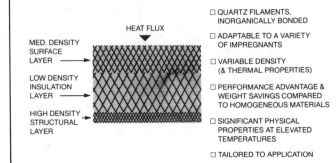

Features of Lock-Heat—A composite material for use in thermal protection applications.

HEAT FLUX

MED. DENSITY SURFACE LAYER

LOW DENSITY INSULATION LAYER

HIGH DENSITY STRUCTURAL LAYER

☐ QUARTZ FILAMENTS, INORGANICALLY BONDED

☐ ADAPTABLE TO A VARIETY OF IMPREGNANTS

☐ VARIABLE DENSITY (& THERMAL PROPERTIES)

☐ PERFORMANCE ADVANTAGE & WEIGHT SAVINGS COMPARED TO HOMOGENEOUS MATERIALS

☐ SIGNIFICANT PHYSICAL PROPERTIES AT ELEVATED TEMPERATURES

☐ TAILORED TO APPLICATION

Typical surface temperature histories on a lifting entry vehicle.

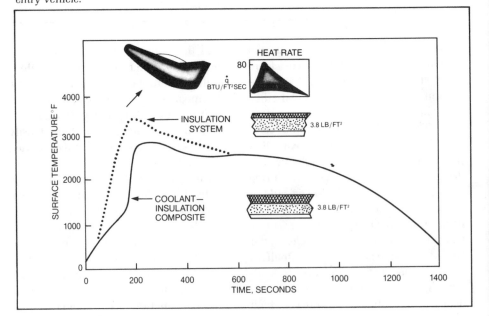

recognition that the insulation belongs on the outside at 2000°F, or as high as ceramics can go. In principle this works to reradiate the energy back out to space rather than take it on board the vehicle and fight the problem from inside. Serendipity. This key insight turned the thermal protection system inside out, and made a cold metal primary structure possible. The concepts looked good enough for LMSC management to fund a significant Space Transportation System design study, which turned into a primitive Star Clipper configuration during 1966 and 1967. At this critical stage the Lockheed-California Company's talented lifting entry aerothermodynamic design team took up residence in Sunnyvale as a corporate resource and the manned space flight charter was thus concentrated at LMSC. The pace accelerated going into the summer of 1968.

As the lifting entry vehicle concepts derived from LMSC's Star Clipper Space Transportation System studies gradually grew to larger sizes, the peak heating rates diminished. Need for the organic impregnant finally disappeared. LI-1500 emerged as the chosen external surface, functioning as a protective reradiative insulation. This turning point marked the beginning of a series of design studies optimized around the LI-1500 formulation and processing approach.

Reradiative Coatings With use of the silica fiber composite as a pure insulator, an undesirable high-temperature emittance characteristic in the critical 0.2 to 3.0μm infrared wavelength range was encountered. The basic concept for optimum heat rejection required high emittance in the significant infrared wavelengths with reradiation of most of the aerodynamic heating to free space.

To retain full reuse capability under space and atmospheric flight conditions, it was necessary to develop a glass surface coating. Requirements for this coating included:

—a match of the thermal expansion with the base composite silica structure

—high thermal emittance at all temperatures in the critical wavelengths below 3μm,

—capability of surviving cyclic heating to 2500°F.

The initial external glass coating developed was based on a borosilicate formulation with a Cr_2O_3 admixture.

First Flight During 1968 NASA flight-tested the LI-1500 pure silica fibrous Reusable Surface Insulation (RSI)

Artist's concept of the pre-Shuttle Lockheed-sponsored Star Clipper stage-and-one-half lifting body configuration ascending from a desert launching base—circa 1968.

Hard learning was achieved from the initial fuselage "green" (chrome coated) tile test panel—decimated in combined thermal-acoustic cyclic exposure at NASA-JSC during December 1971.

Microstructure of early LI-1500 is shown in this scanning electron microscope photograph.

During tests of various thermal protective materials on NASA's Pacemaker reentry vehicle, Lockheed LI-1500 (lower portion of body to right of centerline) was the only one to survive 2300° F reentry temperatures unscathed.

material, along with several candidate ablators, aboard the Pacemaker reentry test vehicle. The reentry vehicle was extensively instrumented, and was recovered after a boost trajectory that simulated the peak heating rates and temperatures anticipated in advanced lifting entry vehicle concepts. Maximum surface temperatures reached 2300°F for the LI-1500 test panel. Inspection after recovery revealed no cracking, melting, or shrinkage of the lightweight material. At this point feasibility of a high-emittance coated-silica reradiative tile system for thermal protection was established. Development forged ahead. The first contract from NASA was awarded in 1969.

OUT OF THE LABORATORY

The significant antecedents of a Space Transportation System had all emerged by late 1968: high pressure hydrogen-

oxygen engines, drop tanks, high L/D delta planform airframes, and a ceramic tile thermal protection system. Now it was time to get down to brass tacks and design it.

Material Requirements During the industry and NASA design studies for a Space Shuttle orbiter, conducted from 1969 through 1973, many alternative *reusable* TPS concepts were evaluated. The ablators used on all previous reentry vehicles were not of course reusable, so they were excluded as possible concepts. The metallic heat shields and active cooling systems considered were too complex and heavy. Thus, it was established that a fiber ceramic tile system offered the lightest weight, highest payload capability, and best potential for meeting the shuttle mission requirement of 100 launch and reentry cycles without significant refurbishment. A number of compet-

ing fibrous composite insulation formulations appeared suitable to enter qualification testing by NASA and the contractors. Discriminating qualification test requirements among these candidates gradually focused on the 100-mission simulation of a reentry heat pulse coupled with a 100-mission launch profile simulation of 165-dB acoustic levels.

Laboratory Process Development Research and development continued in 1969 at LMSC's Palo Alto Research Laboratory as the material requirements became known in cyclic thermal tests. These requirements translate to morphological stability and minimal shrinkage for 15 hours at 2300°F. This is equivalent to 100 missions over a 10-minute peak heating reentry trajectory, which was dictated in turn by concurrent evolution of vehicle configurations in the parallel reentry vehicle engineering design studies. The requisite LI-1500 microstructure thus began to emerge. Investigations at LMSC continued to improve thermal shock and acoustic performance, while reducing density to the LI-900 formulation.

Configuration Evolution Thermophysical properties and design requirements for LI-1500 were derived from intimate TPS installation and mechanical design studies that paralleled the ongoing NASA Space Shuttle Phase B System Definition contracts. At LMSC, the spacecraft and launch vehicle systems integration approach continued along lines of stage-and-one-half configurations with the LS-200-5 lifting body and over-the-shoulder drop tank staging. Characteristically, the aerothermodynamic design requirements of wing loading, leading edge radius, and in-depth TPS thermal response were markedly influenced by the growth in vehicle size driven by NASA's 15 by 60-foot payload bay requirement. Final design iteration of the LS-200-5 late in 1970 outlines the emerging double-delta planforms with body flap and elevon control surfaces around a cluster of hydrogen-oxygen engines.

INTO PRODUCTION

Pilot Production In 1970 LMSC obtained the first of several NASA technology contracts for development of a rigid fiber ceramic TPS. LMSC's confidence in superiority of the silica-based fiber ceramic process was demonstrated by establishment of an LMSC pilot production facility, which was augmented in 1971 to produce both LI-1500 and the lighter weight LI-900 material to support an extensive NASA Space Shuttle TPS test and evaluation program.

Key aspects of process development during this period were process control, and purity and consistency of the silica fiber manufactured by Johns-Manville and refined by LMSC. The amorphous silica fiber produced from high-purity sand is like a super-cooled liquid and has a very low thermal expansion coefficient. Crystalline forms of silica such as quartz or cristobalite have a coefficient over thirty times higher. Transformation from the amorphous structure to the crystalline forms is associated with drastic changes in physical properties as well as totally unacceptable shrinkage and distortion of a sintered fibrous structure.

Much work was done to raise silica-fiber purity to 99.9 percent and reduce alkali impurities to the range of 6 ppm in some instances. Horror stories abound concerning leakage of impurities into the fiber, binder, and sintered product, as well as the coating frits and the critical coating interface. Ultimately, in the transition through a pilot production phase, 140 manufacturing process controls were applied from raw material to finished tile. Dependable performance from lot to lot was achieved in all requisite thermophysical and morphological parameters.

Competition By 1972 the Space Shuttle was in a full-blown competition for the prime system contract. The many alternative system concepts had narrowed down to the solid rocket booster and external tank configuration now familiar as Space Shuttle. As a major contender, LMSC continued with heavily funded in-house component preliminary design effort as well as the NASA funded system definition contract studies. Technical attention turned to refinement of TPS installation design details as well as the total logistics and space operations environment. Areas of analytical design and test included: strain isolation system, cold soak of aluminum primary structure, tile gaps, Room Temperature Vulcanizing (RTV) bond system, filler bar, compound curvature, sidewall coating, gap filler, and many other real world technical constraints. These concepts were well established in time for the competition in 1972.

On 26 July 1972 NASA awarded the Space Shuttle system prime contract to Rockwell International Corporation. Principal focus for LMSC then became the TPS itself. LMSC continued through spring of 1973 with an unbroken string of TPS development contracts with NASA, supporting evolution of the system design and installation concepts, and materials characterization. LI-900, LI-1500, and 0042 coatings were produced in the LMSC pilot production facility. Numerous test articles were manufactured for the expanding TPS

Engineering prototype model of Orbiter forward TPS area, illustrating the installation and strain isolation bonding system on an aluminum honeycomb composite primary structure.

Final evolution of the Lockheed LS-200-5 lifting body stage-and-one-half configuration under NASA Space Shuttle Alternate Concepts contracts in December 1970.

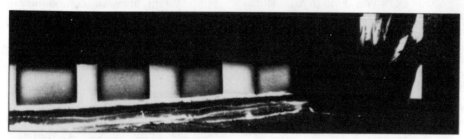

The basic raw material for LI-900 and LI-2200 is a short staple 99.9 percent pure amorphous silica fiber derived from high-quality sand. At top left, premeasured silica fiber is loaded into a carousel hopper. At top right, the silica fiber contents of one of the hoppers is dumped into a chamber where it is mixed with deionized water prior to pouring into a clear plastic mold and forming into a block. At center, the blocks are sintered in furnaces at temperatures up to 2400° F. At lower left, after each block is trimmed and cut into quarters, each quarter is individually machined to precise dimensions to produce a specific tile. After coating and water proofing, the tiles are gathered in arrays that contain an average of 22 single tiles. At lower right, the arrayed tiles are held in place with their coated surface down against the outer mold line of the array frame in the same position they will later assume on the Shuttle. They are then machined as a group along their inner mold line.

competitive evaluation program in NASA research facilities.

Sudden Death With award of the system prime contract to Rockwell International a new phase of competition opened between the silica LI-900 and LI-1500 system developed by Lockheed and the General Electric mullite system baselined in the NASA award. Other competing fiber ceramic material candidates were introduced by McDonnell-Douglas and Martin-Marietta. An extensive series of tests of these competitive materials extended from August through November of 1972.

and were undertaken by NASA to resolve the final selection. The arena was in NASA test facilities at JSC, ARC and KSC, backed up with Battelle and other supporting facilities. The LMSC LI-900 and LI-1500 demonstrated 100 thermal cycles to 2300°F within specified thermal conductivity and backface temperature response requirements. The GE mullite system exhibited excessive backface temperature response and high thermal conductivity, especially in the high-temperature response requirements. Indications of coating cracks and potential substrate fracture appeared

along the way as testing became more severe, and acoustic cycling to simulate ascent dynamic loads was introduced. Testing was then extended to a program of sequential thermal-acoustic cycles simulating the full ascent and reentry environments.

By mid November, LI-1500 had demonstrated the equivalent of 20 sequential thermal-acoustic cycles to 2300°F and 160 dB. A sudden death shoot-out was commissioned in December 1972 with all of the competitive materials in a single large 24-tile array in sequential thermal-acoustic tests at NASA-JSC.

Coating spray area in the tile manufacturing facility at LMSC where the tiles receive a spray coat of reaction cured glass (RCG).

Production units/blocks from which tiles are machined are sintered in this Harper high-temperature elevator kiln.

Some of the odd-shaped tiles that shield NASA's Space Shuttle from fiery temperatures when it returns from space to earth's atmosphere. No two tiles on the Space Shuttle are exactly alike.

Thermal protection units are formed from silica fiber in this casting tower device.

After 20 equivalent cycles, only the LI-900 and LI-1500 tiles remained intact.

LMSC won the recompetition for the TPS subcontract in 1973 with the NASA decision to baseline LI-900 for the Space Shuttle. The proposal as well as the NASA tests demonstrated compelling technical superiority of LI-900 in every conceivable aspect from sequential thermal-acoustic cycle performance to coating integrity in salt and seagull dew. In addition, LI-1500 had achieved a remarkable performance in demonstrating 100 thermal cycles to 2500°F, thermal overshoot to 3000°F, and acoustic overshoot to 174.5 dB.

Subsequently the density of the LI-1500 formulation (15-lb/cu ft) was considered to be too low for use in some specific areas of the Orbiter—such as control surfaces, landing gear door frames, and access panels—which were sensitive to mechanical damage. The higher density LI-2200 formulation (22-lb/cu ft) was introduced for use in these areas, and eventually it was used in lieu of all the LI-1500 applications on the Orbiter.

Full-Scale Production Another transition was initiated in 1973 with preparations for a full-scale production facility at LMSC's main plant in Sunnyvale, Calif. This 43,000-square-foot plant featured the latest in automated fiber-blending and slurry-casting facilities, precision-controlled kilns, and com-

The Space Shuttle Orbiter Columbia about to touch down at Edwards Air Force Base in California with a T-38 chase plane in attendance—the beginning of a new era of space transportation.

period to produce the thousands of complex interlocking shapes required. The vehicle configuration was translated into computer tapes that drive the numerically controlled mills used to machine each individual tile.

The last major hurdle in creating Columbia's thermal protection system was the manufacturing challenge inherent in producing finished machine tiles at LMSC in Sunnyvale, California to fit each of the 30,000 or more specific cavities to form an aerodynamic surface on an aircraft built in Palmdale, California.

puter-controlled inspection equipment. Typically, arduous manufacturing process improvement tasks during scale-up paralleled the challenging technical resolution of material characteristics and performance. The early ground was revisited and several alternate sources of silica fiber evaluated. Early methods of thermophysical property evaluation evolved into sophisticated test methods and analytical modeling of thermal transport mechanisms that took

years to perfect.

The Last Major Hurdle Although the technical challenges of the laboratory-to-production scale-up required much effort during the early years, that task was overshadowed by the effort required to convert Rockwell International master dimension engineering data and coordinates into a finished tile. Numerical control machining techniques and array frame tooling concepts were implemented in the 1977 to 1979 time

POSTSCRIPT

By 1973 the easy work had been done. As the NASA Space Shuttle moved on through design and construction, the number of people involved with TPS grew from hundreds to thousands. A lot of tiles were built that never made it. Engineering and testing went on and on. This is another story. It is extensively documented in the technical literature and the press with full credit to those in NASA, Rockwell International, and LMSC who made it work.

The TPS evolved in complexity and a new generation of materials was cre-

REFERENCES

R. M. Beasley and R. B. Clapper, "Thermal Structural Composites for Aerospace Applications," presented at the 67th Annual Meeting of the American Society for Testing and Materials, June 1964.

R. M. Beasley and Y. D. Izu, "Design and Construction Techniques for Radomes for Superorbital Missions," presented at the OSU-RTS Symposium on Electromagnetic Windows, Columbus, Ohio, June 1964.

W. G. Witte, "Flight Test of High-Density Phenolic-Nylon on a Spacecraft Launched by the Pacemaker Vehicle System," Tech Rept. No. NASA TMS-1910. Langley Research Center, Nov. 1969.

R. M. Beasley, Y. D. Izu, et al. "Fabrication and Improvement of LMSC's All-Silica RSI," in Symposium on Reusable Surface Insulation for Space Shuttle—Vol. I Tech Rept. No. NASA TMX-2719, Nov. 1972.

Symposium on Reusable Surface Insulation for Space Shuttle—Vol. II, Tech Rept. No. NASA TMX-2720, Nov. 1972.

R. P. Banas and G. R. Cunnington, Jr., "Determination of Effective Thermal Conductivity for the Space Shuttle Orbiter's Reusable Surface Insulation," paper No. 74-730 in AIAA/ASME Thermophysics and Heat Transfer Conference, July 1974.

K. J. Forsberg, "Producing the High-Temperature Reusable Surface Insulation for the Thermal Protection System of Space Shuttle," presented at the XIV Congress International Aeronautique, June 1979.

Cooper, Paul A. and Holloway, Paul F.: "The Shuttle Tile Story," Aeronautics and Astronautics, January 1981.

Korb, L. J. and Morant, C. A., et al.: "The Shuttle Orbiter Thermal Protection System"; Ceramic Bulletin, Vol. 60, No. 11, 1981.

Banas, R. P., Elgin, D. R., et al.: "Lessons Learned from the Development and Manufacture of Ceramic Reusable Surface Insulation (RSI) Materials for Space Shuttle Orbiters"; NASA/LARC Conference "Shuttle Performance: Lessons Learned," Langley Research Center, 8–10 March 1983.

ated that ended up on subsequent orbiters or on the shelf against future requirements. This also is another story.

Beyond the lead ship *Columbia* and its sisters in the fleet, there is satisfaction at LMSC in founding a unique sintered fiber ceramic technology that inevitably is spreading to other sectors of engineering application. The technology is now being reinvented in the growing fiber refractory industry that is contributing a new product to the nation's industrial economy.

involved with NASA and NACA in the design and development of advanced turbo-machines, rocket engines, and cryogenic propulsion systems; and at NASA-MSFC was responsible for preliminary design of the Saturn V/Apollo launch vehicle system and the S-IV B stage. Mr. Schramm has served on many government interagency and industry committees and working groups. He was graduated with a B.S. in Mechanical Engineering from Lehigh University.

Flight Research Center, for flight tests designed to measure turbulent heat transfer on the X-15 research airplane.

Mr. Banas received a B.S. degree in Aeronautical Engineering from the University of Illinois in 1960, pursued various graduate courses at USC and UCLA, and has written many technical papers concerning heat transfer in fiber ceramic heat shield materials.

WILSON B. SCHRAMM, Lockheed Missiles & Space Company's Product Development Representative for advanced fiber ceramic composite materials and applications, has been associated with manned spaceflight systems and relevant propulsion and vehicle engineering technologies for most of his career with Lockheed, NASA, and NACA. During the past 21 years with LMSC, Mr. Schramm has been responsible for vehicle engineering and design integration studies of launch vehicles and cryogenic stages, the USAF Manned Orbiting Laboratories program, and in several capacities in program management for the conceptual design and system definition phases of NASA's Space Shuttle. He has also been responsible for development planning and cost analysis of RPV, missile, and munitions systems, and supported various DOD and NASA system acquisition, research, and development proposals and contracts.

Since 1978 Mr. Schramm has managed development and application of new fiber ceramic composite materials for the NASA Space Shuttle as well as other defense and aerospace systems. For 20 years prior to his association with Lockheed, Mr. Schramm was

RONALD P. BANAS is Program Engineering Manager of Thermal Protection Systems in the Space Systems Division of the Lockheed Missiles & Space Co., Inc. Since November 1979, Mr. Banas has had full engineering responsibility for the fiber ceramic High-Temperature Reusable Surface Insulation (HRSI) materials produced by LMSC for the Space Shuttle Orbiters. This assignment also involves planning and directing the development and scale-up to production of new and improved HRSI materials.

During the past 18 years with LMSC, Mr. Banas was on the staff of Engineering Manager for Thermal Protection Systems, responsible for various NASA TPS technology contracts. In 1973 he wrote the material property and environmental tests sections of the successful proposal for the Space Shuttle HRSI subcontract from Rockwell International. In previous LMSC assignments he performed various lifting entry aerothermodynamic heating and insulation sizing calculations, and was involved in all phases of material property tests and environmental testing of HRSI materials.

Prior to joining LMSC in 1965, Mr. Banas was an Aerospace Technologist and Project Leader at NASA/Dryden

Y. DOUGLAS IZU, as a Senior Staff Engineer, directs the developmental efforts of the Advanced Technology Laboratory Materials and Processes Engineering Department. He has conceived, conducted, and directed programs in engineering and manufacturing materials and processes development of advanced organic and inorganic composite material systems for structural, reentry, space environments, mirrors, radomes, and other applications for missile and spacecraft during his twenty years at LMSC. Mr. Izu has been involved in the process development of fiber-reinforced specialized high-performance structures utilizing filament-winding and other related techniques since 1964. He was a key person in the materials development and production implementation of the Lockheed fibrous heat shield and ceramic coating materials for the Space Shuttle Orbiter. He has directed the laboratory effort and served as task leader in composite materials development in support of various government contracted programs. Prior to joining LMSC in 1961, Mr. Izu was with the Boeing Company and the Mare Island Naval Laboratory. He has authored several papers on composite materials and holds a B.S. in Chemical Engineering, 1950, from Stanford University.

CHAPTER 2

MATERIALS

Survey of the U.S. Materials Processing and Manufacturing in the Space Program

Edited by
E.C. McKannan
Material Processing In Space Projects Office
George C. Marshall Space Flight Center

Prepared for Office of Technology Assessment of the Congress

MPS might better be described as materials science and engineering using low gravity. The primary motivation of the program is to use the unique environments of space for scientific and commercial applications. The elimination of the Earth's gravity during the production of common materials affords opportunities for understanding and improving ground-based methods or, where practical and economical, producing select materials in space. Large factories or mills producing huge quantities of materials, as is often the case on Earth, are not expected in space in the near future. Materials that might be produced in space, typically, would be of low-volume but high-value commercial interest. In the more distant future, extraterrestrial materials may be mined and processed for use in space applications.

To promote the potential commercial applications of low-g technology, the program is structured: (1) to analyze the scientific principles of gravitational effects on processes used in the production of common materials, (2) to apply the research toward the technology to control production processes (on Earth or in space, as appropriate), and (3) to establish the legal and managerial framework for commercial ventures.

PRESENT PROGRAM AND STRUCTURE FOR DEVELOPING MATERIALS PROCESSING AND MANUFACTURING TECHNOLOGY IN SPACE

A. Description of Present NASA Materials Science in Space

1. Federal Funded Research

The low-gravity, high-vacuum environment associated with orbiting space vehicles offers unique opportunities to investigate various processes in ways that cannot be duplicated on Earth. Gravity-driven convective stirring, sedimentation, and hydrostatic pressure are virtually eliminated. Materials can be melted, shaped, and solidified in the absence of a container. Processes that require an extremely high vacuum in conjunction with large heat loads or the generation of large quantities of gas may be performed in the wake of an orbiting vehicle.

The low-gravity environment offers a new dimension in process control. Since gravity-driven effects are no longer significant, diffusion-controlled conditions may be easily realized without resorting to the severe restrictions required in Earth's gravity. Many terrestrial processes can be improved or optimized by a better understanding of flows from which better control strategies can be devised.

The elimination of sedimentation in low gravity allows the study of a number of phenomena that cannot be adequately studied terrestrially. The absence of hydrostatic pressure eliminates the tendency for materials to sag and deform when melted and eliminates the need to contain liquids during processing.

Containerless processing eliminates problems such as container contamination and wall effects. Thermal processing and measurements can be done at temperatures well above the melting points of any known material or on materials that are extremely reactive in the melt. Containerless melting in the high vacuum, in the wake of an orbiting vehicle, could produce materials of unprecedented purity.

a. <u>Crystal Growth</u> - Of paramount importance in any crystal growth system is the control of the environment at the growth interface. Compositional and/or thermal fluctuations in the fluid phase (whether it be melt, solution, or vapor) can give rise to inhomogeneities or defects in the growing crystal. Since unstable thermal gradients are virtually impossible to avoid in any growth system, some convective stirring will almost always be present in convectional growth techniques. Such convective stirring is generally thought to be detrimental to control of the growth process and is often considered to be the cause of many growth problems.

(1) Melt Growth - Crystal growth by solidification from the melt is the most widely used technique for production of high technology single-crystalline materials. The success of the technique depends on the control of the composition, temperature, and shape of the solidification interface. This control is often complicated by convection in the melt which affects both the heat and mass transport to the interface. This can cause compositional variations and growth fluctuations.

There is reason to believe, based on Skylab and Apollo-Soyuz experiments, that two major advantages can be realized by growing valuable single crystals in space, i.e., the ability to establish a steady-state, diffusion-controlled boundary layer at the growth interface, and the ability to eliminate growth-rate fluctuations. Flight experiments are being developed to explore how a low-g environment might be used to overcome some of the difficulties in growing crystals. The solidifying surface is difficult to control on Earth and the growth is susceptible to breakdown.

(2) Solution Growth - One advantage of growth from a saturated solution, in which the solute is incorporated into the growing crystal interface is the control it provides over the temperature of growth. This makes it possible to grow crystals that are unstable at their melting points or that exist in several forms depending on their melting points or that exist in several forms depending on their temperature. A second advantage is the control of viscosity, thus permitting substances that tend to form glasses when cooled from their melt to be grown in crystalline form.

A number of interesting systems can be grown from transparent solutions at moderate temperature which allows the detailed study of the growth process and how it is related to growth environment. Because growth from solution requires transport of solute to the growth interface and removal of solvent, it is important to understand how the growth and perfection of the crystal are influenced by this transport process. Since the solvent virtually always has a different density from the solute, solute-driven convection is unavoidable in terrestrial processes. In fact, forced convection or stirring is generally employed in an attempt to maintain a uniform concentration of solution.

(3) Vapor Growth - Particular attention has been given in recent years to the growth of whiskers and to crystalline films from the vapor. Gravity-driven convection can play an important role in the transport of the vapor from the hot source to cold seed. Also, the growth environment in the vicinity of the seed can be influenced by convective effects. There will be both thermal and compositional changes arising from the heat of sublimation released as the vapor deposits. These effects together with natural convection arising from temperature gradients in the

container can be expected to result in nonuniform growth conditions which would affect the perfection of the growing crystal.

Advantages that are expected from the study of vapor growth in space are: the avoidance of deformation of crystals that are weak at their growth temperature, the ability to grow films under unique vacuum conditions, and the possibility of containerless growth.

Several chemical vapor growth experiments were conducted on germanium-selenide (GeSe) and germanium-telluride (GeTe), using iodide (I_2) as the transport gas. The experiments were conducted at various pressures and in the presence of an inert gas in order to investigate the transport rates in low-g environment. In space, substantial improvement in crystalline structure was obtained in terms of fewer defects by visual inspection of micrographs. The most surprising result, however, was in the transport rates. Growth rates were substantially higher than expected, indicating that some unforeseen convective effect is operating in low-g or that some gravity-driven effect is lowering the diffusive transport in the ground-based experiment. In either case, it is apparent that such transport is not as well understood as we previously thought. Additional experiments are being planned to elucidate these transport processes and to explore the possible advantages of growing technically interesting systems such as HgCdTe by this process.

(4) Floating Zone Growth - Floating zone crystal growth is a variation of melt growth in which the melt does not contact a container. This is accomplished by supporting vertical polycrystalline rod at both ends and melting a portion of it with a suitable heater. This floating zone is supported along the sides solely by surface tension. By moving the heater or the rod, the hot zone can be made to move along the axis of the rod, melting the materials ahead of it and growing a crystal behind it. Often the rod is rotated or the two portions are counter-rotated to even out any thermal variations.

The primary advantage of this technique is the absence of wall effects. Many materials of practical interest are highly corrosive in the melt and will partially dissolve any container. Because the electronic properties are dramatically affected by an extremely minute trace of impurity, uncontrolled wall contamination is a serious concern. Also the absence of wall-induced stress during the solidification process can lead to substantial improvement in crystalline perfection. For these reasons, float-zone growth is an extremely important process, especially for producing high-quality silicon and other electronic materials for applications in which high purity and perfection are required.

Floating zone growth is subjected not only to the control and stability problems at the interface that are encountered by ordinary melt growth but has also a new set of problems associated with the free surface. Because the hydrostatic pressure in the molten zone must be supported by surface tension, the length of the zone is limited by the material properties and the diameter of the system. Also the sagging of the melt under the hydrostatic pressure further complicates the control and, consequently, the size and shape of the growing crystal. Thermal convection in the melt is complicated by flows along the free surface of the melt.

It appears that float-zone crystal growth would benefit significantly from the low-g environment in space. The absence of hydrostatic head eliminates the deformation of sag in the molten zone. Zone lengths may be increased. Finally, the use of low-g processing extends floating-zone growth to materials whose low-surface tension prohibits this technique on Earth. One of the major problems in commercially-produced float-zone silicon is the lack of chemical homogeneity. Dopant striations and radial segregation produce considerable variations in electrical properties which can cause problems in very large scale integration devices or in large focal plane arrays. The use of more quiescent growth conditions in space should eliminate this problem.

b. <u>Solidification of Metals and Alloys</u> - Control of the solidification of metals and alloys is the key element in the vast field of metallurgy which provides us with the materials that are fundamental to our high technological society. Gravitational effects, such as buoyancy-driven convection of the melt or the sedimentation of various phases can greatly influence the structure of metals and alloys.

(1) Solidification Kinetics in the Casting of Alloys - In casting of alloys, the fluid, which has a different composition from the solidifying material, often becomes trapped in the spacings between the solid dendrites (tree-like structures extending into the melt). Dendrites break off as growth proceeds. They are carried by density differences or convective flows and either form new growth sites or migrate to the surface and produce mottled surfaces. This process is responsible for producing the fine-grained structure in the interior of a casting, as was recently shown by a series of ground-based experiments using a centrifuge and flight experiments using sounding rockets. This basic information may have application in common products such as iron engine castings.

(2) Unidirectional Solidification - Highly directional properties can be obtained with some castings by directional solidification. This technique affords a degree of control not available in normal castings in that a unidirectional thermal gradient can be imposed and the growth rate can be regulated by moving the sample relative to the thermal gradient. Convective effects can be diminished on the ground but cannot be completely eliminated.

Directional solidification is often used to produce natural composites with reinforcing structure in materials whose components have limited mutual solubility. The size and spacing of the microstructure are determined by the growth rate, the higher growth rates giving rise to finer microstructures. These higher growth rates require very high thermal gradients. Low-g offers several possible advantages. First, the reduced convection in the melt lowers the total heat transfer. Second, shapes can be maintained in low-g with a thin oxide skin, as was demonstrated in experiments performed on German sounding rockets. This allows complicated shapes, such as turbine blades to be melted and directionally resolidified to increase axial strength while using a thin oxide skin to maintain the shape. Without the heavy mold required on Earth to diffuse the heat, a very high thermal gradient can be imposed on the sample which can produce a unique structure.

It has generally been assumed that the microstructure of directionally solidified composites is controlled by diffusion. However, a recent experiment on a rocket flight obtained a finer, more regular microstructure in a maganese-bismuth/bismuth (Mn-Bi/Bi) composite sample than was obtained under identical processing conditions in Earth gravity. This indicates that residual convection may play a more significant role in such processes than was previously suspected.

Another advantage of low-g lies in the processing of off-normal compositions. By eliminating the convective stirring associated with one-g processing, it is possible to build up a layer of the rejected component until the normal composition is reached at the solidifying interface. To provide mass balance, however, it is necessary at steady state to incorporate the same composition into the solid as is present in the bulk melt. This results in both matrix and intrusive material solidifying but in proportions determined by the bulk melt composites. Low gravity provides a new degree of freedom in controlling the size, shape, and spacing of the two phases.

(3) Miscibility Gap Alloys - Hundreds of alloys have been identified which might have useful properties if they could only be processed, but the metals forming the alloys are not miscible on Earth; i.e., they will separate into two liquids at the melt temperature. These alloys are another important class of materials that require reduced gravity for detailed study. Attempts to form alloys or fine dispersions of such material in Earth's gravity by normal freezing are doomed to fail because of this incompatibility. As the melt temperature is lowered into the two-phased region, the low-temperature phase forms solid droplets in the still liquid higher-melting phase. Because these two phases invariably have different densities, they will rapidly separate as the droplets grow. The result is an almost complete stratification of the two metals.

A number of attempts has been made to prepare fine dispersions of the miscibility-gap alloy, aluminum-indium (Al-In), in space. These were generally unsuccessful until the means for controlling capillary flow were demonstrated. When surface tension differences were properly controlled by selection of container materials and reduction of impurities, the predicted fine dispersion was formed during solidification of a hypermonotectic composition in low gravity. The study of miscibility-gap alloys is important from a fundamental and a practical viewpoint; they represent a large class of potentially useful new alloys.

(4) Nucleation in Undercooled Molten Metal - Nucleation and rapid solidification of undercooled melts are also important phenomena that are of fundamental as well as practical interest, in the control of grain structure and, hence, mechanical properties. By containerless melting and solidification, wall-induced nucleation (that is, points on the cold wall where freezing of crystals begins) can be eliminated. This allows the melt to be cooled substantially below the normal freezing point before nucleation eventually occurs. Solidification under these conditions is extremely rapid and can produce unique microstructures and noncrystalline solids, such as palladium-silicon-copper and iron-nickel-phosphorous-boron used in transformer laminations. These materials are thought to be similar to those produced by rapid quenching of thin samples by splat cooling except a thick sample can be processed.

c. <u>Containerless Processing</u> - The MPS program has kindled new interest in containerless processing, i.e., the ability to melt, solidify, or otherwise process a sample without physical contact with a container. There are a number of reasons for wanting to do this: the ability to measure physical properties of high temperature and corrosive materials; to produce ultrapure specimens; to form unique glasses from materials that are reluctant glass formers, either because of their corrosive nature or their tendency to crystallize when they contact the container wall; to study nucleation and the associated time-temperature-transition relations; to study the solidification of deeply undercooled materials; and to fabricate unique shapes without sagging or physical contact, such as highly concentric glass shells for fusion research.

Containerless processing may be carried out on Earth by two basic methods: free-fall facilities and levitation facilities. A free-fall facility, such as the drop tubes at MSFC offers true containerless and near-zero gravity conditions for a very brief time (at most a few seconds). Levitation facilities support the sample (more or less indefinitely) by means of a force applied without solid contact. Such forces may be electrostatic, electromagnetic acoustic, aerodynamic, or hydrostatic. One of the disadvantages of using levitation techniques on Earth is the fact that levitation does employ external forces which may in fact influence the experiment by inducing heating, stirring, or other undesirable effects. Also the applied forces are generally not true body forces but are applied to the surface. Therefore, the sample is still subjected to gravity-driven flows such as sedimentation and natural convection.

Containerless processing in space is essentially a free-fall technique; however, it is necessary to apply low-level levitation forces to compensate for microgravity accelerations and maintain the position of the sample relative to the furnace or experiment chamber. Although such forces may produce some of the extraneous effects encountered in levitation in one-g, the magnitude of such effects can be reduced by several orders of magnitude because of the reduced g-levels.

(1) Nucleation Studies - The elimination of container-induced nucleation has allowed very deep undercooling in excess of 500 C of small (millimeter-sized) droplets of some molten metals such as niobium (Nb) and its alloys. In fact, earlier predictions of the degree of undercooling that was thought to be possible for droplets of this size have been exceeded. Solidification at this degree of undercooling is extremely rapid, comparable to rates obtained by splat cooling of thin samples. Usually nucleation occurs at only one point, and a single crystal results. Small single crystalline silicon (Si) beads have been produced by this method, which may find applications in low-cost solar cells.

Since a melt undercooled to this degree solidifies so rapidly and under highly nonequilibrium conditions, it is possible to form unusual compounds that do not form from equilibrium melting. As mentioned previously, bulk samples of superconductors have been produced in this manner.

One of the major factors in the formation of noncrystalline solids is the time-temperature-transition on cooling. Many substances can be chilled rapidly enough by splat cooling to form a noncrystalline phase, but whether or not such phases can be formed by slow undercooling, which would be necessary to process bulk samples, is a matter of conjecture. In normal solidification, nucleation from the container wall virtually always limits the degree of undercooling that can be achieved. The theory of nucleation is not well-established and is difficult to test. The nucleation and growth of the solid phase in the absence of the container's nucleation sites are obviously an important research area.

(2) Glasses and Glassy Metals - By the elimination of nucleation associated with container walls, it should also be possible to extend the glass forming range of many materials, including some metals. This could result in some new and unique glasses with exotic properties. It may even be possible to delay nucleation long enough for the sample to lose sufficient heat that the latent heat of fusion is no longer capable of raising the material to the normal freezing temperature. The solidification of such undercooled samples is of theroretical as well as practical interest because of the unusual microstructures that may result. Experiments with gallia-calcia galss to control the index of refraction have been initiated.

(3) Containerless Forming - The ability to process an object with very weak constraining forces in the virtual absence of buoyancy forces and hydrostatic pressure offers new techniques for forming materials without physical contact. One process that is of great current interest is a study of the centering mechanisms that operate in the formation of concentric glass shells required for inertial confinement fusion experiments by the Department of Energy. Such shells can presently be made with sufficient precision, in sizes up to several hundred microns, adequate for present research needs. However, in order to provide fusion power on a practical basis, containment shells up to a centimeter in diameter must be produced with a high degree of precision at a low cost. At this time, not enough is known about the mechanisms that are responsible for causing the bubble inside of the glass sphere to center and produce a concentric shell. A better understanding of this process may indicate a method for scaling up the present process to produce larger size shells. On the other hand, if such a process is not feasible on the ground, the shell could be manufactured in space by containerless forming.

(4) Ultrapure Materials - The elimination of crucible contamination is beneficial in the preparation of ultrapure materials. It may be possible, for example, to purify melts by containerless evaporation if the impurities have higher vapor pressure than the host material. Use of the high vacuum associated with the spacecraft wake would be an additional benefit for such a process. Preparation of oxygen-free materials is of technological interest. For example, oxygen impurities in cobalt samarium magnets may be responsible for limiting the magnetic strength of the material. The source of the oxygen is suspected to be the crucible material; therefore, added performance might be derived from containerless melting in an electromagnetic levitator.

A most important application for the preparation of ultrapure material could be the production of ultrapure glass for use in optical wave guides for high frequency communications. Although extremely high purities have been achieved with quartz using the chemical vapor deposition process currently employed for production of optical fibers, the cost is high and the types of glasses that can be produced are limited. With containerless processing, it may be possible to use a broader range of glasses that are less expensive or have different optical properties but tend to pick up trace crucible contaminants because of their chemical activity.

d. Fluids and Chemical Processes - Although the gravitational contribution to chemical reactions is usually negligibly small, many processes are controlled by mass and heat transport which are affected by gravity-driven convection or sedimentation. Often this occurs in complicated ways in which gravity-driven flows are coupled with nongravitational flows such as surface tension-driven convection. Experiments in low-g can simplify such problems by eliminating one set of flows so that the nongravity components may be isolated and studied with a degree of freedom not otherwise possible.

In some processes, it is desirable to keep a component in suspension in order to study nucleation and growth of one phase. By eliminating convection, other transport studies can be made tractable. By eliminating the hydrostatic pressure, free and interfacial surfaces are controlled only by surface tension. This greatly simplifies the study of wetting and spreading and the measurement of contact angles, especially near critical phase transitions where interfacial tension approaches zero. Also, the usually insignificant gravitational contribution to thermodynamic properties is not negligible near a critical phase transition because many terms go to zero. Therefore, there are advantages in studying critical point phenomena in a low-g environment.

(1) Surface Tension Driven Convection - One of the major driving forces for convective flow in the absence of gravity is surface or interfacial tension. There is considerable interest in understanding this phenomenon.

Such flows are expected to play an important role in a number of processes. For example, the thermal migration of bubbles or droplets is driven by temperature variation and surface tension. These flows may be important in getting rid of bubbles in glass or coalescing particles in phase separations. On the other hand, this could also be an unwanted mechanism that causes phase separation in the preparation of some alloys. It is also suspected that surface-tension flows play an important role in flame propagation in flammable liquids. The lower-surface tension near the flame causes fluid to "pull away" from the flame, resulting in a slight depression. The return flow feeds fresh fluid from beneath the surface into the flame.

Perhaps the most compelling reason for studying surface-tension convection is its importance in floating zone crystal growth. This process involves free surfaces and large temperature gradients. The convective transport in a unit gravity field is very complicated. Performing such experiments in a low-g environment provides a method for simplifying the flows by eliminating the buoyancy-driven effect, thus isolating the surface tension effects.

(2) Critical and Interfacial Phenomena - Since the difference between the many thermodynamic properties of two phases vanishes at the critical point, the small contribution from the difference in gravitational potential may become significant. Also since the density of the two phases is virtually always different, gravitational sedimentation will cause rapid phase separation. This limits the time over which the transition can be observed on Earth.

As systems that have two immiscible liquid phases approach the transition to a single liquid phase, the interfacial tension goes to zero. Theory predicts that one of the phases becomes perfectly wetting at the critical point. This is difficult to observe in the laboratory. The existence of one phase that becomes perfectly wetting may be the key to the rapid phase separation observed when attempting to form certain alloys.

(3) Polymerization Phenomena - There are a variety of processes important to industrial applications that may be studied in a low-g environment for the purpose of understanding the role of gravity effects or for taking advantage of the virtual lack of buoyancy-driven convection, sedimentation, or hydrostatic pressure in order to achieve better process control. For example, seeded polymerization of latex may be used to grow monodispersed spheres in a size range that is virtually not achievable by ground-based techniques because of creaming or sedimentation effects. Such spheres would be useful as standards for calibrating counters, sizing membranes, and possibly other biomedical applications.

The absence of gravitational forces provides an opportunity to study the role of chemical fining (bubble removal in glass) by eliminating buoyancy motion. Chemical fining agents certainly will become more important in the future as energy costs continue to rise. Because glass is viscous, velocities of bubbles are extremely low, and very long, high temperature soaks are required. Chemical finers are known that help reduce the time, but the mechanism is not well understood. It is difficult to contain a bubble in a small container for study on Earth.

Another endeavor that takes advantage of the absence of sedimentation is the study of Ostwald ripening used for strengthening alloys. This is a process by which second-phase particles evolve into a distribution of sizes with the larger particles growing at the expense of smaller particles. This effect is quite important in metallurgy where the size of a precipitated second phase is altered by heat treating. Although the phenomenon of Ostwald ripening has been known for a number of years, there is no satisfactory agreement between theory and experiment because of the difficulty of maintaining stable suspensions of the precipitating phase on Earth.

(4) Electrochemical Deposition - Electrochemical deposition is a process that is influenced considerably by gravity because of internal heating of the bath as well as convection from concentration gradients. Yet little work has been done to understand the role of gravity-driven flows. In some cases it is desirable to incorporate inert particles of a second phase into an electroformed product to improve its properties. Maintaining such particles in a suspension and incorporating them uniformly into the structure is difficult in the laboratory. Low-gravity electroplating experiments have been initiated.

(5) Biomedical Applications - A number of processes of interest to the biomedical community that are adversely influenced by gravity effects have been identified. These include electrophoresis, isoelectric focusing, phase partitioning, suspension cell culturing, crystallization of macromolecules, and the study of blood flow. Experiments performed in space can provide valuable insight into the control of the processes on Earth, produce research quantities of unique products, and eventually develop unique separations or products on a preparative scale. A joint endeavor with an industrial firm is evidence of the potential economical benefits that can come from bioprocessing in space.

(a) Separation Processes - Separation of complex material mixtures into their component parts is a goal of extreme importance in many fields but is of particular importance in the various biomedical fields. Indeed, the lack of satisfactory techniques for such separations is now in many cases the limiting factor that impedes further research progress. The kinds of biological components in need of improved separation methods include cells, cell components, and macromolecules.

Every organ of the mammalian body is made up of a great variety of functional cell types, which can frequently be best studied in homogeneous populations of a single cell type. However, cells that differ importantly in their functions are often so similar in their size, density, and appearance that it is impossible to obtain a pure population of any one cell type by conventional separation methods (density gradient centrifugation, for example). In addition to facilitating basic research into normal and abnormal cell physiology, improved cell separation methods could conceivably be used to provide purified cell populations for transplantation into individuals lacking normal cells of that type. Also purified populations of cells that secrete valuable products could be most useful in the commercial production of these substances, because cell culturing processes could be greatly simplified and their efficiency increased by selecting only active, producing cells for culture. The kidney cell separation experiments on Apollo-Soyuz provided a step in that direction.

Purifying macromolecules that have been synthesized is another application of bioprocessing separations technology. Some macromolecules must be laboriously purified from heterogeneous medium such as blood, urine, or tissue. Other macromolecules can be obtained through cell culture, recombinant-DNA technology, or artificial peptide synthesis. In each of these cases, however, the final product is not pure but rather a complex mixture requiring extensive purification. In addition, neither recombinant-DNA technology nor laboratory peptide synthesis can be performed until the desired product has been obtained in strictly pure form in sufficient quantity to allow determination of its molecular structure. For many important macromolecules (e.g., interferon, alpha-1-antitrypsin), this has been impossible to achieve in spite of the many techniques presently available. Methods currently employed for the purification of complex mixtures of macromolecules include ultracentrifugation, filtration and chromatographic (preferential adsorption) methods.

One important fact to consider is that most current cell separation techniques are based either on very nonspecific characteristics of the cell (e.g., density) or on characteristics so highly specific that the method can be used only on previously purified populations of cells. A purified cell population must first have been obtained. Because of these limitations, existing techniques are either quite nonspecific, or else they are "circular" in that it is necessary to possess a purified population before one is able to obtain a purified population. In contrast, both continuous-flow electrophoresis and phase partitioning avoid this liability; with either technique, it is possible to begin with a mixed population of cells about which one knows essentially nothing and to sort them into categories which are specific as well as biologically meaningful. However, both of these methods appear to be significantly impeded by gravity-related effects. Therefore, free-flow electrophoresis and phase partitioning could conceivably provide unique advantages over other techniques if such separations were carried out without the perturbations introduced by Earth's gravity.

1. Electrophoresis - Electrophoresis is a well known technique separation of proteins and other macromolecules on an analytical scale. Such molecules acquire a specific charge when immersed in a buffer solution. An applied electric field interacts with this charge and produces a force which moves the molecule against drag encountered in the medium. This allows materials with different mobilities to be separated.

On Earth, electrophoresis is usually carried out in the presence of a gel to prevent convective mixing. This restricts the quantities that can be separated and precludes use of the technique for cells. One technique for circumventing this restriction is the use of free-column electrophoresis, using vertical columns with a density gradient to stabilize the fluid motion. In this system, the cells migrate against gravity. Although this technique has enjoyed only limited success, it was a first choice for space where density gradients are not required.

Experiments performed on Apollo-Soyuz showed some indication that kidney cells could be separated according to function. One fraction of cells, when cultured, showed a significant enhancement in the production of the valuable enzyme urokinase. Column electrophoresis can be a very valuable tool for space use for obtaining extremely high resolution separations of research quantities of material.

Another means of obtaining electrophoretic separation of cells or macromolecules is the use of continuous-flow electrophoresis (CFE).

By its nature, free-flow electrophoresis is associated with significant problems, many of which are gravity determined. The passage of the electric current causes heating which tends to produce unwanted thermal convection, impairing the resolution of the separation. Sedimenation of the sample material limits the concentration of the sample and thus the throughput. The CFE instruments currently in use are stabilized by using a very thin flow channel to suppress convective flows. This limits throughput and resolution because the sample stream is subject to distortion from wall effects. To determine the benefits that may be obtained by operating in space as well as the most advantageous design for such devices, it was necessary to study the fluid dynamics of electrokinetic separation. It has now been found that most of these difficulties can be avoided in a low-g environment where the buoyant forces would be dramatically lessened. There may be other effects that limit the performance in low-g, but it does appear that there is a good rationale for a flight experiment to explore the effects that are overshadowed by convective problems on the ground.

2. Isoelectric Focusing - In isoelectric focusing, a pH (H + ion concentration) gradient is set up by an applied electric field upon electrolytes added to the buffer solution. The material to be separated is driven by the electric field to a region of pH known as the isoelectric point, where the electrophoretic mobility of the sample is zero. Very high resolutions are possible because the boundaries of the sample bands are self-sharpening. This is a particular advantage when dealing with proteins or other macromolecules because the focusing process counteracts diffusion. Isoelectric focusing is subject to gravity-induced convection and sedimentation problems.

A novel isoelectric focusing machine was recently developed as a potential space experiment. This machine is a recirculating device with a number of fluid loops that come together in a common chamber. The electrolytes migrate through membranes in the chamber to form a stepped pH gradient in the various flow channels. Once the pH gradient has stabilized, the biological sample is introduced and each component migrates to the channel corresponding to its isoelectric point. A ten-channel machine has been built and demonstrated to be capable of separating a number of test materials. It is believed that the membranes could be eliminated entirely by performing the separation in low-g.

The device has attracted considerable attention in the biomedical community, and a number of researchers have delivered samples to be separated. It has been suggested that high-resolution, high-throughput, continuous-flow isoelectric focusing would be a useful method for purifying natural or synthesized products such as polypeptide hormones, interferon, recombinant-DNA products and other macromolecules.

(b) Blood Rheology - Blood is a fluid with a viscosity that is strongly dependent on shear rate. This fact is due primarily to the presence of the red cells. The rheology (study of flow) is further complicated by the fact that red cells form aggregates, of varying sizes in different physiological conditions. Coagulation may occur in addition to rapid sedimentation of aggregates. Rheological considerations appear to be important in several disease states, including cardiovascular disease, diabetes, sickle-cell anemia, and some forms of kidney disease. For example, sedimentation of red blood cells is more pronounced in a variety of pathological conditions, and it forms the basis for clinical tests. One question is raised as to whether the cellular aggregations in pathological conditions seen in the laboratory produce a different pressure drop across a blood vessel. Simultaneous sedimentation precludes an answer to this question. An increase in the pressure drop merits appropriate countermeasures, whereas a decreased pressure drop would constitute a beneficial situation. Obtaining a thorough understanding of the rheology of blood is extremely important. However, the examination of biological cell dispersions in laboratory viscometers is rendered problematic under terrestrial conditions by sedimentation. Research in a reduced-gravity environment to alleviate these problems has been proposed.

e. Extraterrestrial Materials Processing

(1) Goals - The long range goal of space materials science is to provide for the cost-effective use of extraterrestrial materials resources for both space systems and terrestrial applications. This goal requires the demonstration of advanced autonomous and teleoperated machinery for remote operations. The machinery must have a high degree of self-sufficiency and self-replication. The remote operations include exploration, acquisition, staging, conversion, manufacturing, assembly and construction of facilities. The goal also requires demonstration of space-adapted techniques for acquiring and converting raw materials occurring in space into useful forms for space and terrestrial applications.

(2) Development - Development of technology along several parallel fronts is required to accomplish the goals. Exploration of the moon, asteroids and planets will locate the needed mineral supply in terms of distribution and concentration and characterize it in terms of chemical and physical properties. Development of technology in space systems will provide the materials, structures, machine intelligence, robotics, energy systems and transportation systems to reach the goal. Plans for coordination of activities leading to a phased development of a system demonstration are being defined. The first phase will involve project planning, resource evaluation, materials analysis and materials processing experiments. The second phase will involve breadboard demonstrations on individual system components, such as an Earth demonstration of an automatic reducing cell to process lunar or asteroid material into aluminum and/or silicon, and an Earth orbital demonstration of elements of the space materials processing system for concept verification. The third phase will involve designing and building actual space systems hardware, testing and mission operations. A steering committee is currently working on the details of this effort.

(3) Status - Currently, hardware components are being developed in biomedical applications, spacecraft remote deployment and retrieval of experiment packages, and by adaptation of remote manipulators from Surveyor and Viking planetary spacecraft. In the area of intelligence, very large scale integrated circuits are being applied with increased hierarchy to control industrial processes. These can be adapted, directly. Materials processing techniques are being examined for adaptivity to space systems. Finally, for planning purposes, self-replicating algorithms are being formulated.

(4) Plans - Planning is being done to provide for automation of the deployment and operation of free-flying materials processing payloads. Computer control of spaceflight payloads is being decentralized to enhance flexibility, changeout, learning capabilities and independent decision making on individual payloads. Research is directed toward a better understanding of materials processes, including scaling laws and nonlinear phenomena and to develop the best space processing methods for extraterrestrial materials. System engineering is being employed for analysis of competing scenarios.

MPS VARIOUS WORKING GROUPS

MAIN FUNCTIONAL RESPONSIBILITY	WORKING GROUPS		
	DISCIPLINE	SCIENCE	INVESTIGATOR
SCIENCE	IDENTIFY NEW AREAS AND CONCEPTS	ESTABLISH EXPERIMENT STRATEGIES TEST NEW EXPERIMENT CONCEPTS	REVIEW FLIGHT READINESS OF SCIENCE PERFORM PRE-FLIGHT AND POST-FLIGHT ANALYSES
HARDWARE	NO FUNCTION	IDENTIFY LONG-RANGE PLANNING NEEDS PROVIDE REQUIREMENTS FOR CONCEPTUAL DEFINITIONS IDENTIFY TECHNOLOGY DEVELOPMENT NEEDS	PERFORM GCEL TESTS ASSIST IN HARDWARE DEVELOPMENT & TESTING ASSIST IN HARDWARE REFLIGHT REFURBISHMENTS & MODIFICATIONS
MISSION	NO FUNCTION	REVIEW PROGRESS OF CURRENT EXPERIMENT PROGRAM EXAMINE OVERALL MISSION CONFIGURATIONS SUPPLY STRAWMAN REQUIREMENTS FOR DEFINITION ACTIVITIES	IDENTIFY SPECIFIC MISSION REQUIREMENTS TRAIN PAYLOAD SPECIALISTS AND ASSIST IN OPERATIONS

FIGURE 8.

37

III. FUTURE POTENTIAL OF MPS

A. Scope of the Program

Only those aspects of materials processing that combine deleterious effects due to a sensitivity to gravitational force and potential economic feasibility are of interest at this time. The technical aspects of low gravity that are of interest to industrial and scientific materials processing are:

1. Reduction of sedimentation and buoyancy to enable control of multiphase systems and the preparation of variable density solids.

2. Reduction of density gradient-driven convection (such as thermal and solutal convection) to enable direct assessment of the convection effects experienced in industry and research on Earth in crystal growth, solidification, chemical, and biological separation processes, and the subsequent preparation of materials with more control over structure, composition and external surface features.

3. Reduction of hydrostatic deformation in liquids and semisolids enhance industrial and research activities such as floating zones, crystal growth, and solidification, study of critical point phenomena, preparation of high molecular weight crystals, and diffusion of glasses at temperature above the softening temperature.

4. Development of containerless methods for positioning, measuring, preparing and forming molten, reactive materials.

B. Terrestrial Payoffs From MPS Research and Development

In general, the MPS program is interested in studies of process parameters to enhance process control and productivity in Earth processing, in the preparation of limited quantities of precursor materials to provide baseline or reference data, and in the development of methods unique to the space environment by which materials can be prepared which are not possible on Earth. All these interests rely upon the range of process parameters being extended through the reduction of gravity. For example, electromagnetic levitation of some molten metals is possible on Earth; however, in space where heating and levitation can be decoupled, any molten conductive sample can be positioned and heated electromagnetically over a wide range of temperatures. In a second example, glass shells for fusion targets are currently produced up to several hundred micron size in drop towers (free-fall); scale-up to 7 to 10 mm (the projected optimum diameter) may be more easily accomplished in Earth orbit where longer times are possible at high temperature. In a third example, growth of solid-solutal crystals required a combination of high temperature gradients at the growth interface, but low interface migration velocities that are impossible to achieve because of the thermal and solutal convection effects; growth rates and temperature gradients can be decoupled in low gravity, thus, extending the range of stable growth parameters. In all these examples, the first priority is to develop an understanding of the gravitational limitations and to develop theoretical models for one-g and zero-g behavior, followed by breadboarding simple hardware and acquiring preliminary data before committing to major hardware activities. The emphasis is on the information developed from these experiments being a significant contribution to our body of knowledge in materials processing and the technology transfusable to the private sector. The best measure of these activities is the number of technical papers published in the scientific literature. As of August 1980, from 61 investigations active at that time, 106 technical papers were generated over a two-year period. Also, there was much industrial interest. The publication rate is probably comparable to that of similar National Science Foundation (NSF) funded work, but no attempt was made to develop comparisons. Detailed science reviews of the progress in each investigation are carried out on (roughly) an annual basis by a peer group of non-NASA scientists who write anonymous reports. These reports are aggregated and synopsized by the program scientists and fed back to the PI's. This peer review process complements the initial proposal review which is done by mail in the same manner used by NSF. The industrial interest is evidenced by the ground-based investigations by private industry, IGI participation on selected processing ventures. Several such arrangements have been established through the Commercial Applications Office at MSFC. Other less formal industrial liasions have been established through academic institutions such as the Materials Processing Center at MIT.

The following list of priorities represents the thrust of the MPS program and its relation to typical areas of private and public center interest:

1. Crystal Growth Processes

a. Melt growth is the most widely used technique for production of high technology, single-crystal materials for semiconductor chips used in large scale integrated circuits for communications and computers. The MPS program emphasis is concentrated on achieving chemical homogeneity, hence, maximum electrical performance, in HgCdTe and lead-tin-telluride (PbSnTe) semiconductors. These crystals are among the most sensitive and important infrared sensors and most difficult to grow materials on Earth. The two materials bridge the spectrum of growth conditions. In the case of PbSnTe, one component is less dense than the bulk melt, hence the system is subject to solute instabilities. The HgCd Te, on the other hand, has the opposite problem. One component is more dense than the bulk melt. Therefore, it is subject to solidifying interface-shape instabilities. Low-g experiments will determine how well such systems can be grown in the absence of gravity. These are two examples of several commercially valuable crystals whose properties may be enhanced by melt growth in a low-g environment.

b. Float zone growth is a variation of melt growth in which the material can be melted without the deleterious contact with any container wall. Floating zone techniques are widely used to produce crystals such as doped silicon for semiconductors and solar cells. While large, efficient crystals are grown commercially with this process, gravity does limit the size and type of crystals that can be grown and does introduce growth rate fluctuations that cause chemical inhomogeneities that necessitate cutting the crystal into small chips for high performance applications. The MPS program emphasis is on establishing uniform growth conditions in commercially important materials such as indium-doped silicon and CdTe which is a semiconductor with a very high theoretical maximum energy conversion efficiency.

c. Solution growth is an important alternative to melt growth for materials that are unstable at their melting point because the crystals can be processed at much lower temperatures. The MPS program emphasis is directed toward triglycine sulphate (TGS) a room temperature, infrared detector material and gallium-arsenide (GaAs) one of the most important semiconductors for a wide range of applications from microwave devices, to computers, and solid state lasers. TGS is grown from a transparent, water base solution that permits observation of the growth process; furthermore, the infrared detectivity of the currently available material is constrained to about 20 percent of the theoretical limit because of gravity influenced growth defects. Thus, this system represents a good model material that is well characterized and has the possibility of a technological breakthrough if substantial improvement can be realized.

GaAs is one of the most important semiconductors with users ranging from microwave devices to solid state lasers. It can be readily grown in bulk form but with considerable imperfections. Usually, in device fabrication, a thin film of high quality GaAs with precisely controlled additives is grown by liquid-phase epitaxy (layered growth) over a bulk melt grown crystal substrate. However, two problems arise. First, in the growth of the epitaxial layer, the solvent is less dense than the Ga/GaAs solution. Therefore, the growth system cannot be stabilized against convection. Second, it is very difficult to control the saturation at the growth interface by lowering the temperature of the substrate.

Although it is possible to bury the defects in the substrate, the buried defects tend to migrate with time and eventually emerge at the surface, causing premature device degradation or failure. Therefore, for certain applications, it would be highly desirable to have better substrate material. For these reasons the growth of GaAs in low-g is of interest from a theoretical as well as a technological point of view.

d. Vapor growth does not compete favorably with other growth techniques on Earth where large crystals are required because gravity disrupts the vapor transport mechanisms; it is a useful process for growing "whiskers" or thin monocrystalline films and for materials that do not lend themselves to other convenient techniques. The absence of gravity opens new possibilities for the growth of large, flat, pure crystals by the vapor technique; therefore, the MPS program includes the investigation of HgI_2 nuclear detector crystals and HgCdTe and copper-indium-antimony (CuInSb) solid solution semiconductor crystals.

HgI_2 which is an excellent prospect for nuclear radiation detector that can be used at ambient temperature. One of the factors that is believed to limit the performance of this material is its high density and extreme weakness at the growth temperature. Because the crystal has a layer structure, self-deformation during growth under one-g is believed to be an important factor in producing dislocations which degrade the performance as a nuclear energy detector. The growth of such a crystal in low-g could eliminate such strains at the growth temperature. It is also anticipated that the perfection of the crystal might benefit from the more quiescent growth conditions expected in space. The solid solution semiconductor materials are being pursued not only because of interest noted earlier in the materials, but also because earlier flight experiments indicated much higher growth rates in zero-g than those produced on Earth.

2. Solidification Processes

Directional solidification is a casting process used to produce single crystals and two-phase composite materials wherein the microstructure is aligned in a particular direction such that the mechanical and physical properties differ along various axes, or wherein fine, homogenious dispersions are achieved. Common examples of two- (or multi) phase composites might be fiberglass wherein glass filaments are suspended (either unidirectionally or randomly) in a plastic matrix to increase strength and provide anisotropic properties and dispersion hardened steel wherein small carbide particles are included in the steel matrix to improve strength. The advantage of this process is that the growth rate and, hence, the microstructure and properties can be controlled. In zero-g, the tendencies for the second-phase materials, to be uncontrollably mixed through convection or separated by sedimentation/buoyancy, are eliminated and sharper thermal gradients and lower growth rates are achievable. The MPS interest in directionally aligned composites is built upon the extraordinarily high mangetic coercivity measured in space-grown composites of Mn-Bi/Bi. Additional interest is based on the potential of approaching the theoretical maximum magnetic strength of materials such as samarium-cobalt ($SmCo_5$) which is 10 times higher than currently realized on Earth.

The second aspect of directional solidification finds application in miscibility gap alloys that defy preparation in one-g in bulk quantities because gravity-driven effects cause the materials to segregate upon solidification. There are some 500 such combinations of materials that have a liquid phase miscibility gap. If producible, such materials might have such diverse applications as electrical contacts (as replacements for silver and gold) and self-lubricating bearings. Experiments in low-g have successfully produced finely dispersed, homogeneous mixtures of Ga-Bi and Al-In. Other materials, such as Cu-Pb, Cd-Ga, Ag-Ni, Al-In-Sn, Cu-Pb-Al, Cd-Ga-Al, and transparent model materials, are being studied in the MPS to define nongravity segregation phenomena and to establish the techniques to produce these unique materials for property evaluation. While applications may be speculated from theoretical considerations, the inability to produce bulk quantities on Earth necessitates the making of samples in one-g to verify those expectations and the application viability.

Undercooled solidification is the rapid quenching of molten materials at temperatures well below their freezing points. This process is valuable in the preparation of amorphous (glass or glass-like) materials as well as pure single crystal and metastable phases. Materials have a natural tendency to form uniform crystalline structures; therefore, to make a glassy material, the atoms must be "frozen" in a random order before the crystalline state is achieved. The common glass materials can be chilled and solidified by common Earth-based techniques. Other materials can be solidified in the amorphous state, except that the atomic mobility is generally so rapid that heat (especially for molten materials that must be held in a crucible) cannot be extracted before crystal nucleation takes place. Through elimination of the gravitational requirement for a container, a vast array of materials can be processed in the amorphous state in low-g, thus, extending the materials properties available to mechanical and optical designers. The so-called "reluctant glass forming" materials are the basis of the MPS program effort. NASA is doing and sponsoring work on glasses such as CaO-GaO to produce such materials in bulk quantities and to confirm their improved energy transfer efficiency for applications such as laser hosts. The NASA MPS program has developed unique ground-based free fall facilities at MSFC in which extreme undercooling (hundreds of degrees centigrade in excell of existent theory) has been achieved in the production of bulk quantities of pure single crystals and superconducting metastable phases; these materials have not been made in bulk quantities by other methods. The emphasis in undercooled solidification centers on the formation of pure Nb and the superconducting phase Nb_3Ge, which has a high superconducting transition temperature (23.2°K) and offers great promise for electrical transmission and electrical devices. By producing these compounds in sample's large enough to be analyzed by neutron diffraction, the relationship between perfection of the crystalline structures and the superconducting performance can be analyzed. Such studies may shed some light on methods for obtaining superconductors which work at higher temperatures.

Casting technology has already been advanced through zero-g

experiments that unequivocally illustrate how the crystalline structure grows and is subsequently disrupted by gravitational effects leading to the pervasive occurrence of property variations from top to bottom of a casting (an engine block, for example) and shrinkage. Castings made in zero-g have uniform microstructure and properties. Therefore, the MPS program is using the zero-g environment to study the formation and resultant properties of various cast materials (both simple model materials and commercial alloys) to establish process controls and techniques that might be adapted on Earth. Furthermore, NASA is studying the prospects of casting complex, single crystal shapes (such as turbine blades) in space to achieve theoretical maximum properties from the materials.

3. Containerless Processing

Levitation technology is being pursued to develop devices for positioning, melting, manipulating, and resolidifying materials in space without the constraint of containers or crucibles. In space, liquid materials will remain in a stable, spherical drop without containers; thus, small restraining forces are sufficient to keep the drop where desired. The processing of materials without the necessity of containers is an exciting and unique capability of the space environment and permits the formation of pure materials without contamination from the container, permits the formation of amorphous (glass) materials that cannot be made on Earth, and permits the measurement of physical properties of molten materials at temperatures that exceed the melting point of crucibles needed on Earth. The MPS program technology is directed toward the development of high temperature acoustic levitators (or positioning devices) for use with materials that can be processed in a gaseous environment, electromagnetic levitators for use with electrically conductive materials, in either gaseous or vacuum processing environments, and electrostatic levitators for use with dielectric materials that need to be processed in vacuum environments. Low-g flight experiments have been conducted successfully with both acoustic and electromagnetic devices, and the practical application of this technology to a vast spectrum of both scientific and commercial processes can be realized through the elimination of detrimental gravitational effects.

4. Biological Separation Processes

Bioseparation technology is being addressed because Earth-based techniques for producing high purity materials in significant quantities from complex biological mixtures are adversely affected by gravity. In the gravity-free environment of space, separation techniques that are based on electric fields and biological material surface characteristics become highly efficient. Furthermore, such separation techniques are inherently gentle and do not damage or destroy live cells or other material. The focus of the MPS program is in developing the technology for separation techniques such as electrophoresis, isoelectric focusing, and phase partitioning. Substantive advances have already been made in the improvement of the Earth-based technology through the MPS program, and a venture by a private sector firm has been formalized to explore the viability of commercialization of pharmaceuticals separated by such techniques in space.

5. Fluids Mechanics and Chemical Processes

Fluid mechanics are critical to nearly all material processes since at some point in the process, the materials exist in either the liquid or gaseous state and are, therefore, subject to gravitational disturbances. The MPS program has undertaken to analyze the processes and to develop appropriate theoretical and mathematical models for both the one-g and low-g aspects once such understanding is imperative to understanding the Earth-based property limits and the viability of low-g experimentation. The development of adequate mathematical models (at least for simple materials) is especially important since many, if not most, commercial material processes have been developed empirically over long periods of time (dating as far back as the Bronze Age) and often involve such complex mixtures and combinations of materials that they defy analysis of the reactions and interactions taking place. Low-g offers an opportunity to isolate one of the major variables in understanding these processes.

Chemical processes are being studied to elucidate the effects of gravity in processes where particle size and geometry may affect the chemical reaction kinetics. Currently, the MPS program is investigating the reaction size limitations in producing uniform, microscopic particles for applications such as blood cell counter and electron microscopic calibration, calibration of pore sizes in living or other membranes, and for tagging biological materials. Under one-g, as the particle size is increased, they tend to aggregate and sediment. An early low-g flight experiment may provide valuable information on chemical process controls applicable to the field of polymer chemistry.

In conclusion, the materials processes that can most probably be improved by operations in low gravity have been identified. They are contained solidification, including single crystal growth and polycrystalline solidification of metals and alloys; containerless solidification of crystalline and new amorphous materials; and new bioseparation techniques, among other fluid and chemical processes. Research has been initiated in each of these disciplines and some flight experiments have been defined.

ACRONYMS

AN	-	Applications Notice
AO	-	Announcements of Opportunity
AR&DA	-	Advanced Research and Development Activity
ASES	-	Advanced Solidification Experiment System
ASTP	-	Apollo-Soyuz Test Project
ATD	-	Authority to Proceed
BIO	-	Bioseparation
CFE	-	Continuous Flow Electrophoresis
EMC	-	Electromagnetic Containerless
ESC	-	Electrostatic Containerless
FES	-	Fluids Experiment System
FL	-	Float Zone
GCEL	-	Ground Control Experiment Laboratory
HGDS	-	High Gradient Directional Solidification
IGI	-	Industrial Guest Investigator
JEA	-	Joint Endeavor Agreement
JPL	-	Jet Propulsion Laboratory
JSC	-	Johnson Space Center
KSC	-	Kennedy Space Center
LaRC	-	Langley Research Center
LeRC	-	Lewis Research Center
MAUS	-	Materialwissenschaftliche Autonome Experimente Unter Schwerlosigkeit
MDAC	-	McDonnell Douglas Astronautics Company
MEA	-	Materials Experiment Assembly
MEC	-	Materials Experiment Carrier
MLR	-	Monodisperse Latex Reactor
MIT	-	Massachusetts Institute of Technology
MPS	-	Materials Processing in Space
MSFC	-	Marshall Space Flight Center
NBS	-	National Bureau of Standards
NSF	-	National Science Foundation
OSS	-	Office of Space Science

ACRONYMS (Continued)

OSTA	-	Office of Space and Terrestrial Applications
PI	-	Principal Investigator
P/L	-	Payload
PS	-	Power System
SES	-	Solidification Experiments System
SPAR	-	Space Processing Applications Rocket
SRT	-	Supporting Research and Technology
STAMPS	-	Scientific & Technological Aspects of Materials Processing in Space
STS	-	Space Transportation System
SWG	-	Science Working Group
TEA	-	Technical Exchange Agreement
TEXUS	-	Technologische Experimente Unter Schwerelosigkeit
VCG	-	Vapor Crystal Growth

STS-2 Landing

LOCKHEED MISSILES & SPACE COMPANY

FACT SHEET

SPACE SHUTTLE HIGH-TEMPERATURE REUSABLE SURFACE INSULATION (HRSI)

LI-900/LI-2200

When NASA's space shuttle orbiters reenter earth's atmosphere from space, they encounter temperatures as high as 2300°F. To protect the orbiters and the astronauts from such intense heat, Lockheed Missiles & Space Company has developed an ultrapure silica fiber insulation that transfers heat so slowly, a piece of the material can be held by the edges with a bare hand only seconds after being removed from an extremely hot oven. In those few seconds, heat has rapidly escaped from the surface of the material, but fiery interior heat has not yet moved to the outside to replace it.

Conversely, the black reflective coating applied to silica "tiles" allows ninety percent of the searing heat generated by reentry to be radiated back into the atmosphere. Several hours are needed for the heat trapped inside the material to work its way out through the silica fibers. Consequently, temperatures on the shuttle's aluminum skin never exceed the structural design limit of 350°F.

The unique material is manufactured in two forms, LI-900 (Lockheed Insulation, 9 pounds per cubic foot) and LI-2200 (22 pounds per cubic foot). Since LI-2200 is denser and stronger than LI-900, it is used to cover access doors, the main landing gear door and other areas of the orbiters that require tougher insulating material. Together, the two types of material cover 70 percent of the exterior of Columbia and Challenger the first two vehicles of the shuttle orbiter fleet.

The basic raw material for the reusable Lockheed insulation is a 99.7% pure silica fiber that is derived from high-quality sand.

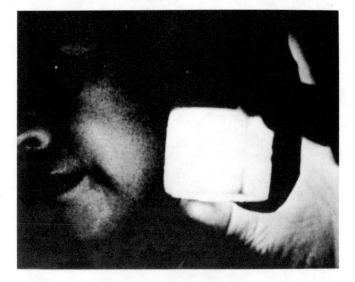

White hot at 2300°F, the glow from a cube of HRSI held in the technician's bare hand provides the only illumination in this photo.

Lockheed obtains the material from Manville Products Corporation, Denver, Colorado. The insulation made from the fiber was selected by NASA for the shuttle because of its light weight, low thermal expansion and high-temperature stability.

SECOND GENERATION

Beginning with Discovery, the third shuttle, LI-2200 will be replaced with a new material that will reduce each orbiter's weight by about 1000 pounds. Called FRCI-12 (Fibrous Refractory Composite Insulation), the material weighs 12 pounds per cubic foot, and is just as strong as the heavier LI-2200. Here's why: FRCI-12 is composed of 80 percent pure silica fibers and 20 percent of a 3M Company fiber called Nextel. Nextel contains a small amount of boron that welds the pure silica fibers and Nextel fibers into a rigid structure during high-temperature sintering in a furnace. The result is a lighter material with strength equal to the denser and more loosely structured LI-2200. Potential weight reduction will allow future orbiters to carry more payload or additional passengers. Columbia and Challenger each contain about 3000 LI-2200 tiles.

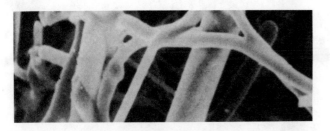

FRCI @ 2420°F

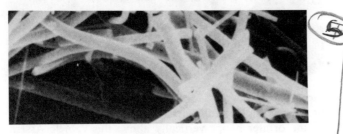

LI @ 2350°F

Fiber Fusion Characteristics

Display of HRSI tiles shows some unusual shapes needed on various parts of the shuttle.

WHY TILES?

The shuttle is far from the solid, rigid craft it appears to be. The aerodynamic pressures and vibrations experienced during launch cause the orbiter to bend and twist substantially. In addition, the contraction experienced in the cold temperatures of space and the subsequent expansion in the heat of reentry, cause significant flexing of the vehicle. For these reasons, designers decided to use tiles instead of larger panels that would have been more rigid and more likely to crack. In order to isolate the tiles from the orbiter's movements, Rockwell International workers bond them to a synthetic felt called a strain isolation pad (SIP) which in turn is bonded to the shuttle's skin. Each orbiter requires nearly 31,000 tiles - each a different size, shape and thickness.

TILE PRODUCTION PROCESS

The primary difference in manufacturing the pure silica tiles and FRCI tiles is in mixing the materials that form the slurry. The other steps are essentially the same.

First, a slurry mixture of fibers and deionized water is released into a plastic container and formed into soft blocks.

TILE PRODUCTION PROCESS (continued)

After a colloidal solution is added, the material is dried in a microwave oven and sintered in a large furnace at 2300°F. The rigid blocks, measuring typically 15" x 15" x 6-1/2", are cut into quarters and then machined to precise dimensions.

Coatings of silica frit are next baked on in 15-mil (0.015-inch) thickness. Tiles for the underside of the shuttle and other areas exposed to temperatures up to 2300°F receive a black reaction cured glass borosilicate glass coating (RCG). For lower temperature use at 750° - 1200°F on upper surfaces, a white silica compound is used with shiny alumina oxide added to better reflect the sun's rays and keep the shuttle cool on orbit. The tiles are also treated with a silane waterproofing polymer to prevent them from absorbing moisture from rain or humidity.

SHUTTLE INSULATION REQUIREMENTS

The economic feasibility of a reusable Space Transportation System hinges on protecting the vehicle from reentry heat so that it does not require significant refurbishment between flights. In the past, manned spacecraft have used only ablative heat shields that are destroyed during reentry.

As early as 1957 Lockheed began investigating a broad range of candidate shuttle insulating materials including zirconium compound, alumina and aluminum silicates. After 1961, the work concentrated on finding a suitable all-silica material. An earlier, but heavier version of LI-900 was successfully tested in 1968 during the reentry of NASA's Pacemaker spacecraft where surface temperatures reached 2300°F.

Today, the successful flights of the shuttle have proven that coated and water-proofed thermal tiles meet all mission requirements imposed by a reusable space transportation system.

The unique combination of physical properties, obtained only with fused silica fiber, assures the tiles' ability to:

* Keep temperatures of the aluminum airframe within structural design limits
* Remain reusable for 100 missions
* Meet special heat-resistant needs in addition to those met in standard ascent and reentry trajectories caused by on-orbit thermal cycling, the plume from the shuttle's rocket engines, emergency orbiter abort procedures and pre-entry heating during "once-around" flights near the upper atmosphere prior to landing
* Provide an insulated surface that offers good aerodynamic smoothness

Most of the nearly 31,000 silica tiles that protect NASA's Space Shuttle Columbia from fiery reentry heat are in view as the orbiter hangs vertically above the Vehicle Assembly Building floor at the Kennedy Space Center, Florida. (NASA photo)

SHUTTLE INSULATION REQ'TS (cont.)

- Minimize absorption of rain or humidity during periods when the shuttle is on the ground between missions

HOW MANY TILES WILL BE NEEDED?

Lockheed is presently under contract to Rockwell International, space shuttle prime contractor, for four complete shipsets of tiles. Excluding additional tiles that were manufactured during the research and development phase of the program and those that may be needed as spares, this means that Lockheed is providing more than 100,000 individual tiles for the first four spacecraft. Those tiles will consume approximately 6,500 cubic feet of raw material and would cover more than an acre of land.

PLACEMENT ON THE SHUTTLE

The tiles are attached to the shuttle by Rockwell International. Finished tiles are bonded to a strain-isolator pad of Nomex felt and then to the shuttle's aluminum skin with a room-temperature adhesive.

The drawing below shows the locations of various insulating materials on the Orbiter Columbia and Challenger. LI-900 white tiles are used in areas that reach 750°F to 1200°F temperatures on the sides of the tail section, engines, fuselage and on top near the front of the wings. Black high-temperature tiles cover the entire underside and some small areas on the top side that may reach temperatures from 1200°F to 2300°F. Together, the tiles cover about 70 percent of the spacecraft. In addition, reinforced carbon-carbon is used on the ultra-hot nose and leading edges of the wings and nomex felt in regions heated to less than 750°F on cargo bay doors and some upper wing surfaces.

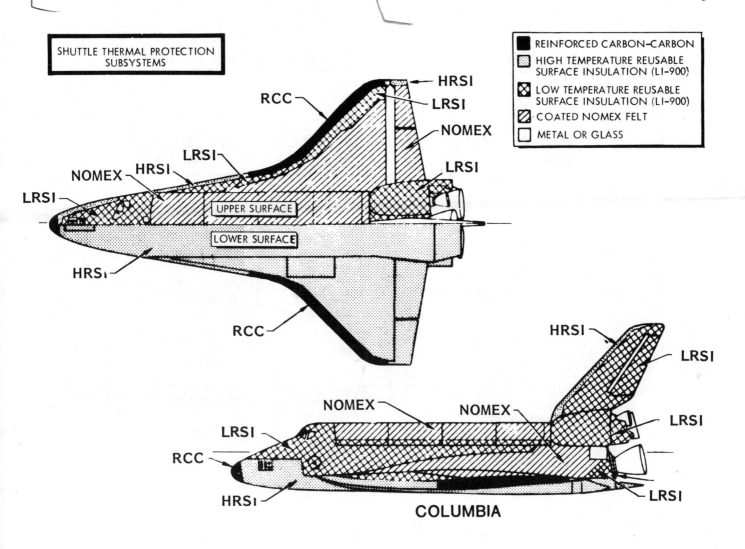

SHUTTLE THERMAL PROTECTION SUBSYSTEMS

- ■ REINFORCED CARBON-CARBON
- ▒ HIGH TEMPERATURE REUSABLE SURFACE INSULATION (LI-900)
- ▨ LOW TEMPERATURE REUSABLE SURFACE INSULATION (LI-900)
- ▤ COATED NOMEX FELT
- □ METAL OR GLASS

COLUMBIA

CHALLENGE OF ASSEMBLY

To save weight and assure a perfect fit, computer-controlled milling machines are used to match precisely the curvature of each tile's underside to the contour of the shuttle's skin at the exact point the tile is to be bonded. For that reason, no two tiles are exactly alike.

A code number on each tile indicates where it is to be placed. This means the job of fitting the tiles together on the spacecraft is as tricky as assembling a jigsaw puzzle twice the size of a basketball court. To further complicate matters, the small gaps between the tiles must be uniform to within ±20 mil – a very tiny margin for error.

To simplify the milling and assembly procedures, the tiles, after upper surfaces are individually machined, coated and waterproofed - are gathered in arrays that contain an average of 22 single tiles. A typical array frame without tiles is shown below (at left). There are more than 1,100 different array configurations for each shipset of tiles. The tiles are placed in the frame with their top (coated) surfaces down against the outer mold line (OML) in the same relative positions they will later assume on the shuttle orbiter. Locked into place, they are then machined as a group along the inner mold line (IML). Still held in their fixed locations by the frame, they are next bonded on the IML side to Nomex felt strain-isolator pads, and, in turn, to their designated spot on the shuttle's aluminum surface. The diagram below (at right) shows the order in which layers of insulation pad and bonding agent are applied.

ARRAY FRAME

INSULATION PLACEMENT

FOR MILLING AND INSTALLING TILES

ON SHUTTLE'S UNDERSIDE

ON SHUTTLE'S UPPER SIDE

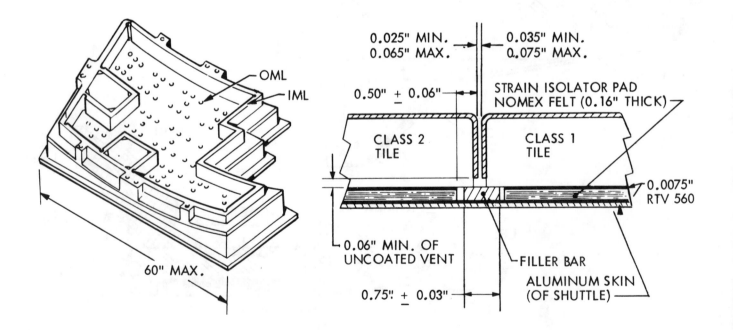

OML

IML

60" MAX.

0.025" MIN.
0.065" MAX.

0.035" MIN.
0.075" MAX.

0.50" ± 0.06"

STRAIN ISOLATOR PAD
NOMEX FELT (0.16" THICK)

CLASS 2
TILE

CLASS 1
TILE

0.0075"
RTV 560

0.06" MIN. OF
UNCOATED VENT

FILLER BAR

ALUMINUM SKIN
(OF SHUTTLE)

0.75" ± 0.03"

A4

THE DELIVERY SCHEDULE

Orbiter 101 - the first shuttle vehicle that rolled out of Rockwell International's Palmdale, California plant in September 1976 - was fitted with about 200 LI-900 tiles for test purposes. It did not require a complete shipset of tiles since early flight tests were conducted at low altitudes and did not involve a fiery atmospheric reentry. These initial flight tests performed in 1977 verified the feasibility of the tile attachment concept.

The second spacecraft, Columbia, required the full protection of Lockheed insulation as it was the first shuttle vehicle tested and flown on orbit. Lockheed began delivering tiles for Columbia in early 1977 and deliveries were essentially completed in February 1979. The company has a contract to produce tiles for the other three shuttle vehicles and will deliver any spares required to replace damaged tiles.

STEP-BY-STEP HRSI-MANUFACTURING PROCESS

Lockheed's 40,000 sq.-ft. manufacturing facility features the very latest in furnace, coating, machining, inspection/test equipment, and numerical or computer control systems. Quality control and precision are the key considerations at every stage of the manufacturing process shown below and on the next page. During peak periods of production - which began in early 1977 - these processes, along with array fabrication, employed more than 400 people.

(1) At left: Premeasured silica fiber is put in turntable "bucket" enroute to mixing chamber

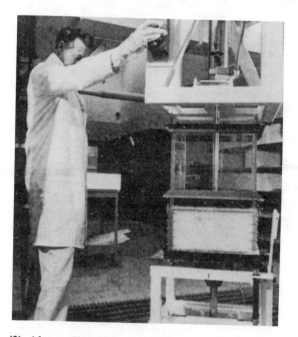

(3) Above: Water is squeezed out to form block; colloidal silica binder is then added

(2) At left: Mixer drains slurry of fiber and deionized water into clear plastic mold

(4) Above: After drying in a microwave oven, blocks next enter this large sintering furnace

(7) Above: X-rays and computerized systems verify density and dimensions of every tile

(5) Above: Blocks are cut into quarters from which final tiles will be machined

(8) Above: The tile is placed in a masking structure which holds it secure during the coating process

(6) Above: Quarters are individually machined to precise dimensions

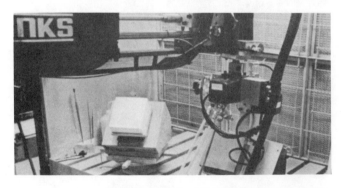

(9) Above: Automated sprayer adds either high (black) or low (white) temperature coating

In the final steps, coatings are baked on in the sintering furnace and a waterproofing polymer is applied by vapor deposition in a vacuum chamber.

Materials Engineering in the Sky

Processing in space could create a whole new world of materials

By John C. Bittence, Senior Editor

Some time, probably before 1982, the Space Shuttle will deliver the first scientific payload into orbit. Once in place, the Skylab will begin humming as dozens of experiments, scientific and military, public and private, are activated. Somewhere deep within the Shuttle, in a five-cubic-foot container, a movie camera will whir while a small cloud of moisture begins to crystallize into a snowflake. The first of hundreds of materials-science experiments destined for Space Shuttle delivery and pickup is taking place. It's a study to observe the growth of a snowflake under gravity-free conditions.

In later flights, scientists hope to produce superpure glass, inexpensive solar cells, and a host of metallic alloys that don't even exist on earth. By 1984, between 100 and 120 materials experiments per year will be conducted in space. Eventually, these experiments may turn into full-fledged manufacturing operations to produce earth-bound products. Ultimately — though certainly not in the near future — raw materials for these operations may be mined and imported from the moon.

The attraction of 0g

Why will we go thousands of miles above the earth to process materials and make parts? Primarily, because the space environment offers a combination of conditions that are difficult, if not impossible, to duplicate on earth. First of all, space offers a cheap solar-energy supply. In addition, near-perfect vacuum (down to 10^{-14} torr) is right at hand, temperature extremes between -200 and $+200°F$ (-130 and $+93°C$) are readily available, and the environment is always free from atmospheric-related corrosives and impurities.

The most valuable condition in space, however, is the almost total absence of gravitational force. Although a trace, or microgravity (up to $10g^{-4}$), is always present, this value is practically zero for most materials applications.

Near-zero gravity does not result from the satellite being beyond the earth's pull, because it is not. Although gravitational forces in orbit are much less, the earth's pull on a satellite is still significant. Rather, low gravity, or weightlessness, results from the balance of forces that keep the satellite and its contents in orbit. Microgravitational forces arise when this delicate balance is upset by simple movements of the craft caused by orbit corrections, equipment operations, or even motion of the occupants. "We are always dealing with microgravity, never zero gravity," points out Frank J. Jelinek, associate section manager, Battelle-Columbus Laboratories. "But for most purposes, we regard it as 'zero.'"

Weightless conditions in orbit may be a bonanza to metals researchers. At "zero g," everything floats, so molten materials can be solidified in vacuum without ever contacting a container wall. In addition, at zero g, mixtures of light and heavy substances stay mixed. Materials solidified at zero g are homogeneous; there's no settling out of more dense constituents, and no turbulence during freezing.

One of the first phenomena researchers want to study in space is the solidification of immiscible metals — metals that normally don't form alloys on earth because one is pulled out of solution by gravity before the molten material solidifies. Over 500 such binary metallic systems with miscibility gaps on earth have been identified. Scientists are eager to determine whether any of these materials are superconductors, superplastic, or possess any unusual magnetic or electrical properties when solidified in zero gravity. "The potential here is tremendous," observes Battelle's Alan J. Markworth, a physicist anxious to get the Space Shuttle under way. "Although we've primarily been studying the aluminum-indium system (because it's easy to work with), there are hundreds of alloys that could develop into something useful," Markworth adds.

Miracles from immiscibles

Battelle has been studying miscibility-gap metals since 1973, but only recently have they been able to test out their theories under low-gravity conditions. Alloys of Al-40In and Al-70In, cast during a few precious seconds of near-zero gravity during the free-fall trajectory portion of SPAR II rocket flights from Marshall Space Flight Center, yielded some suprising results.

First, in the absence of gravity, lesser known physical phenomena become extremely important. Surface tension, for example, plays only a minor role in metal casting and solidification on earth, where gravity is the dominant force. In space, however, in the absence of gravity, surface tension plays an active part.

In addition, thermal viscosity and thermal conductivity in the cooling

Flying foundries that shape with sound waves

One of the virtues of working with materials in space is that they are weightless. Highly reactive or ultrapure materials can be melted while floating between induction coils or acoustic transducers without being contaminated by a container or die. In the absence of gravity, surface tension becomes the prevailing force, so the molten mass takes the shape of a sphere.

NASA proposes using acoustic energy, or sound waves, to shape these molten spheres into more useful forms. In a system devised by the Jet Propulsion Laboratory at CalTech, the molten metal is retained in a container by an energy field set up by six transducers. By adjusting acoustic energy emitted from any of the transducers, engineers could contour the metal into a variety of geometrical shapes. Or, as illustrated, the metal can be pulled out of the container. Surface tension shapes the withdrawn metal into a rod or fiber, but two additional transducers can be used to flatten the molten metal into a sheet or ribbon. At this point the metal is allowed to solidify.

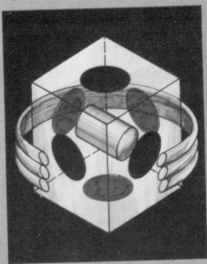

Solid raw material floats weightlessly, constrained by energy from six transducers.

metal can become bothersome. In one Battelle rocket experiment reported by Markworth, thermal gradients, rather than gravity, induced segregation — and immiscibility — in an Al-In casting. One theory suggests that thermocapillary flow (the Marangoni Effect) is the culprit leading to segregation in the absence of gravity.

Gravity-driven convection currents also are absent from the molten metal in space. The result is a finer microstructure than is possible on earth. In some cases, unexpected and unidentified phases have turned up in low-gravity castings. Paradoxically, and totally opposite from earth-bound metal behavior, slowly cooled metal in space has a finer microstructure than rapidly cooled samples.

Scientists admit that they are just beginning to understand the behavior of immiscible metals. In fact, most of the experiments to date have raised more questions than answers. Ultimately, scientists agree that space experimentation will contribute substantially to our technical understanding of materials behavior, including solidification dynamics, vaporization and condensation, and diffusion.

Specific metallurgical achievements

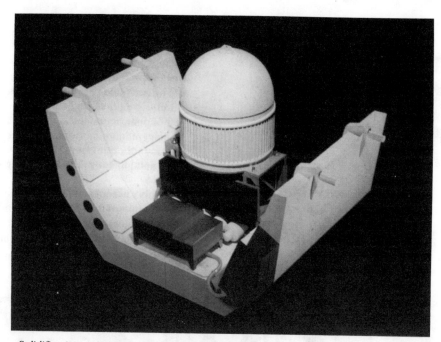

Solidification-experiment portion of TRW's MPS project will be installed in the Space Shuttle cargo bay. Domed section houses a carousel with about two dozen material samples that will be inserted into the high-temperature furnace below. Five gas bottles below furnace hold helium for cooling. Gold-covered structure houses electronics. The entire experiment will be preprogrammed and automated, although manual override is possible.

that scientists hope will result from zero-gravity experimentation include manufacture of superhigh-strength magnets in space for use on earth, production of extremely strong, fine-grained welds between metals not weldable on earth, continuous casting of dispersion-strengthened metals, melt-

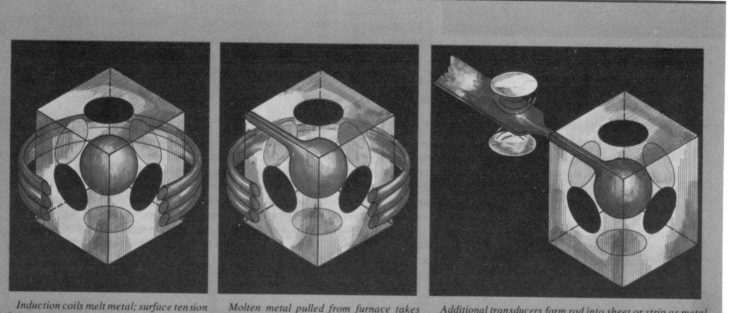

Induction coils melt metal; surface tension forms it into a sphere.

Molten metal pulled from furnace takes shape of rod, fiber or whisker.

Additional transducers form rod into sheet or strip as metal is allowed to solidfy.

A pinch of basalt

Although it's still a generation away, researchers are already predicting the day when we will be mining the moon for raw materials. Most likely, these materials will be delivered to orbiting factories that manufacture huge structures in space. A 13-mi² satellite to collect solar power for the earth, for instance, could be made of 90% lunar materials.

Lunar mining is attractive because it takes 22 times more energy to send material into orbit from the earth than from the moon. In addition, a wealth of metals and minerals are available right on the surface of the moon, as the table shows.

Notably absent from the lunar crust is hydrogen, thus making the production of water impossible. Moon mining and beneficiation will have to rely on anhydrous techniques. Also, because of low gravity and lack of an earth-like atmosphere, mining will probably be unmanned.

The space program is often maligned as a waste of money, but knowledge gained from missions such as this Apollo 17 moon sampling has been extremely valuable in improving our understanding of materials and processing.

Materials the moon holds in store for us

Element	Earth rank	Earth composition, ppm by weight	Moon composition, ppm by weight Highlands	Mare
Oxygen	1	466,000	417,000	446,000
Silicon	2	277,000	212,000	210,000
Aluminum	3	81,300	69,700	133,000
Iron	4	50,000	132,000	48,700
Calcium	5	36,300	78,800	106,800
Sodium	6	28,300	2900	3100
Potassium	7	25,900	1100	800
Magnesium	8	20,900	57,600	45,500
Titanium	9	4400	31,000	3100
Hydrogen	10	1400	54	56
Phosphorous	11	1050	660	500
Manganese	12	950	1700	675
Carbon	17	200	100	100

Data courtesy Convair Div. General Dynamics.

ing and casting of highly reactive metals without the need for crucibles or molds, production of ultrapure metals, and the development of new families of inter-metallics and immiscible alloys.

Superfibers, superglass
Engineering fibers on earth made from glass, carbon, or ceramics have only a fraction of their theoretical strength. Surface irregularities, grain boundaries, and crystalline defects, all introduced during manufacture, reduce fiber strength and ductility. Space-produced fibers, on the other hand, probably will more closely approach theoretical properties.

Techniques have been proposed for growing high-strength, monocrystalline fibers, or whiskers, of materials such as silicon carbide to be used as reinforcement for metals and composites. In an early, rocket-borne experiment with silver, whisker-strengthened with 5% SiC, bending strength was improved by 20% and deflection under load improved by a factor of seven.

Even more impressive, however, are predictions for improvements in optical glass and solar silicon made in space. Some of the first materials that will be studied in the Space Shuttle are ultra-pure laser glass and fibers for optical wave guides.

Battelle has developed a new technique for producing glass and glass-ceramics that is especially adaptable to the special conditions of space. Called the Sol-Gel process, the scheme requires both the zero g and high vacuum of outer space (ME 7/80).

In the realm of glass and ceramics, researchers expect another materials bonanza. "Magnetic" glass ($FeBO_3$, containing 70% ferromagnetic iron) has great potential for magneto-optical devices. Unfortunately, this unusual material cannot be made on earth in a form suitable for these devices. It can, however, be produced by containerless melting in space.

Glass products with dispersion, refractive, and other properties unavailable in earth-made materials possibly can be created in space. For instance, lasers can be made more efficient by increasing the calcium content of the Nd-Ca-Li-Al silicate glass optics. Unfortunately, this can't be done on earth, since contact by calcium with the container walls during melting causes devitrification. In space, however, high-efficiency laser optics can be made by containerless melting. According to an Owens-Illinois study, improvement in laser efficiency should be well worth the cost of processing in space.

Similar predictions are made for space-processed solar silicon and various semiconductor materials such as Ga-As. According to some estimates, high yield from space-processed semiconductor materials will easily offset the high costs of space processing. The added benefits of space processing will stem from improved reliability of space-made semiconductors.

Typical glass and ceramic products

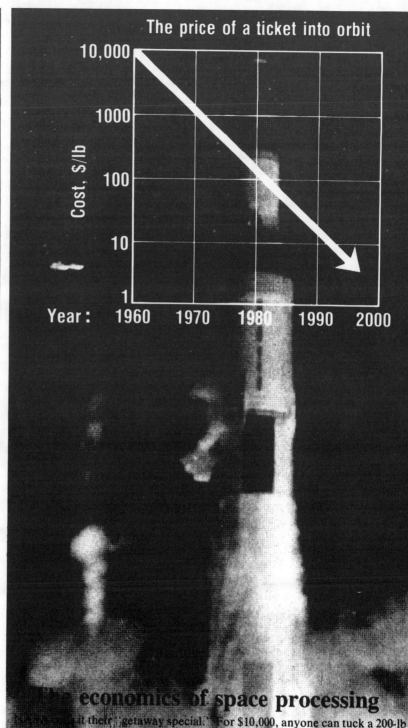

The price of a ticket into orbit

The economics of space processing

NASA calls it their "getaway special." For $10,000, anyone can tuck a 200-lb payload into the bowls of the Space Shuttle for delivery — along with the rest of the cargo — into orbit around the earth. An additional $10,000 of the cost is government subsidized, so the total cost for a trip into space is about $100/lb, or 100 times less than the cost in 1960. By 1990, a trip into orbit should cost $25/lb — still an expensive ride.

On the other hand, efficiency of solar-energy generators can be at least doubled if critical silicon cells were made in microgravity space. The rejection rate for certain earth-grown, high-purity semiconductor wafers can be reduced from present-day high rates to practically "zero" when they are grown in orbit. And an ultrapure version of a biological material called collagen produced in space would be almost priceless for medical treatment of burns, replacing blood vessels, and producing artificial corneas.

The cost benefits of producing certain materials in space, especially in minifactories that can be refurbished and returned to orbit, are indeed worth the steep transportation costs.

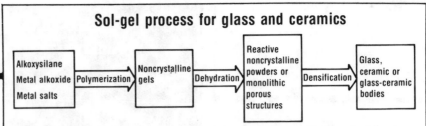

Sol-gel process for glass and ceramics

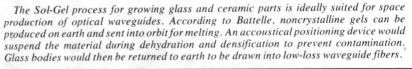

The Sol-Gel process for growing glass and ceramic parts is ideally suited for space production of optical waveguides. According to Battelle, noncrystalline gels can be produced on earth and sent into orbit for melting. An accoustical positioning device would suspend the material during dehydration and densification to prevent contamination. Glass bodies would then be returned to earth to be drawn into low-loss waveguide fibers.

that scientists hope to develop from space experimentation include "selective-wavelength" optics, holographic storage devices, better integrated circuits, ferroelectrics, and high-performance lenses and mirrors.

MPS ready to fly

In 1978, NASA recognized the potential benefits from materials experimentation in space by awarding the Materials Processing in Space program (MPS) to TRW Defense and Space Systems Group. The assignment was to develop a new facility for materials R&D in space with the ultimate possibility of performing commercial work in orbit. According to NASA, the MPS facility will be used to "conduct applied research in fluids technology and materials solidification relating to electronic materials, metals, ceramics, glass and chemicals." Manufacturing processes outlined as being particularly adaptable to the space environment include melting and casting, vapor deposition and powder metallurgy.

Two MPS payloads emerged. The first, the manned facility for fluids experiments and vapor crystal growth, will be installed in one of the Spacelabs. The second unit, a pallet-mounted experimentation system, will ride in the Shuttle's cargo bay. This unit is designed to perform more than two dozen solidification experiments in a high-temperature furnace by preprogrammed control.

By about 1984, the NASA power module for free-flying orbital missions should be ready, and TRW's MPS program will be ready for its next step — launching of the Materials Experimentation Carrier (MEC) which will link up with the power module for missions of up to 120 days. By 1987, MEC will become a manned, fully serviceable, continuously orbiting facility, called the Materials Experimental Module, MEM-1. It will be used for advanced R&D materials and pilot-plant production. By 1990, MEM-1 will be replaced

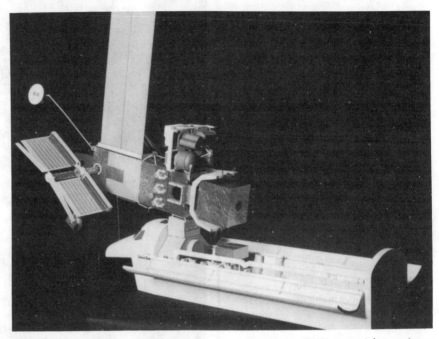

When NASA's free-flying 25 kW power system goes into orbit, TRW's materials experiment carrier (MEC) will go along. As shown here, the MEC is being docked on the power system before both are detached from the Shuttle. MEC, which will succeed MPS in 1984, will be designed for either 120-day missions or continual orbiting. Prototype and pilot production experiments will be carried out in MEC.

by MEM-2, whose specific design and mission are yet to be established.

The mine in the moon

Some researchers envision the day when a facility similar to the MEMs would be supplied with raw material from the moon, or perhaps from nearby mineral-rich asteroids, although beneficiation of these elements will require novel technology. Unlike earthbound beneficiation, which relies on water, acids, bases and an atmosphere, moon processing of raw materials probably would be based on electrostatic beneficiation powered by solar energy. Because of low gravity and lack of an

earth-like atmosphere, mining on the moon would probably be totally automated.

The rationale for moon mining stems from the lower escape velocity — approximately one-fifth that of the earth — required to remove a mass of material from the surface. A NASA-funded study by General Dynamics, Convair Div., proposes that materials be catapulted from the moon's surface and captured by space "garbage trucks" for batch delivery to the production facility for processing. Ultimately, NASA would consider building entire space stations in space, deriving 90% of the raw material for construction from the

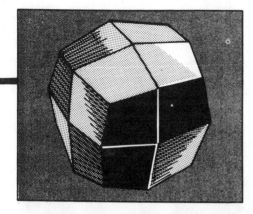

Vapor deposition is well suited for space. In fact, it's been taking place extraterrestrially since the beginning of time. Moon samples have intrigued researchers because certain materials take on new and unexpected morphologies when vapor deposited under conditions of low gravity. This previously unknown form of Fe, a variation of a 24-sided trisoctrahedron, is virtually a pure and perfect crystal. What can its properties be?

"Win $5000 and a trip to the U.S. by suggesting the best idea for an experiment to be performed aboard the Space Shuttle," said this Tokyo newspaper ad. The winning idea, growing a snowflake in gravity-free environment, was suggested by two teenagers. Battelle liked the idea because it will help us understand how gravity affects solidification. The newspaper is paying for development and Shuttle costs.

moon. Lunar material would cut costs of in-orbit construction by about one-third.

When will all this take place? "These things are *all* conceivable," points out Battelle's Jelinek. "Probably by the time your children and my children are senior citizens, they will be a reality." General Dynamics researchers agree, observing that current and projected capabilities to produce solar energy sufficient to carry out lunar mining will not even exist by 1990.

A National Research Council committee, studying NASA's original MPS proposal in 1978, was even more cautious. "The committee has not discov-

ered any examples of economically justifiable processes for producing materials in space," they reported. "When gravity has an adverse effect on a process, stratagems for dealing with it can usually be found on earth that are much easier and less expensive than recourse to space flight."

On the other hand, Europe, Japan, and the USSR have given top priority to materials science experimentation in space. In a General Accounting Office review last winter of space-processing programs proposed or carried out by NASA so far, the U.S. Congress was taken to task for its low funding and support of materials experimentation in

space. In fact, according to GAO, foreigners could easily assume the lead in materials science in space, since materials experiments in early Shuttle/Spacelab missions will be predominantly performed by other nations.

The European Space Agency, ESA, currently developing the first Spacelabs in cooperation with a consortium of European businesses, has selected materials processing as one of the two disciplines to be included in the first flight. ESA has a list of about 120 materials science experiments they would like to propose for the first and subsequent Spacelabs.

Begins with a snowflake

From manned factories in orbit to unmanned strip mining of the moon . . . it will all start by growing a tiny snowflake aboard the Space Shuttle. The snowflake will be grown because two Japanese 16-year-olds suggested the experiment, and Battelle was commissioned to include it in one of the first of two Space-Shuttle payloads for it which it purchased flight options. The teenagers won a contest sponsored by a Tokyo newspaper, *Asahi Shimbun,* out of more than 16,000 entries proposing ideas for research experiments that could be conducted aboard the NASA shuttle.

Toshio Ogawa and Yasuhiko Oda wanted to see whether snowflakes grown in the weightlessness of space differed from those grown under gravitational influence of the earth. Battelle's Alan Markworth liked the idea; he's anxious to determine whether the basic platelet flake will grow in three dimensions. Differences in morphology between earth and space-grown flakes will be evaluated for scientific significance. "Although," concedes Markworth, "a lead-off experiment such as this will have a lot of public appeal, too. Especially in Japan." Cost of the snowflake project: $100,000, which includes a $10,000 ticket for the ride in the Shuttle. The sole product of the experiment will be a length of four-color film. ∎

Shuttle facelift
could be in metal

Multiwall metal panels now in development may some day replace the ceramic insulating tiles that have been protecting the Space Shuttle from the intense heat of re-entering the earth's atmosphere. The new system, conceived by L. Robert Jackson, a structures engineer at NASA's Langley Research Center (Hampton Bays, Va), can provide thermal protection at temperatures ranging from 700 to 2700F, up to 3500F with an assist from heat-pipe cooling.

The panels comprise alternating flat and dimpled sheets joined at the dimple crests as shown in the drawing below. Depending on the degree of thermal protection required, the panels may be all-titanium or combinations of titanium and more-heat-resistant materials. Superplastic forming is a likely choice for forming the dimpled titanium sheets, which would then be diffusion-bonded to the facing.

Heat conduction through the panels is minimal because the layers touch only at the dimple crests, a contact area of less than 0.2% of the surface area. The long conduction path and low heat conductivity of titanium contribute to the thermal barrier.

The multiple layers, each of which is a radiation barrier, inhibit radiative heat transfer. Keeping the cells small prevents heat transfer by convection. Evacuating and sealing the cells eliminates conduction by gaseous products.

The all-titanium system would be limited to missions encountering temperatures in the 700-900F range. For temperatures of 900-1600F, a metallic enclosure supports a fibrous insulating filling, pressure being supported by either an outer foil-gage superalloy and dimple-core sandwich or a superalloy honeycomb sandwich joined by beaded edge seals. For temperatures of 1600-2700F, either ordinary or dispersion-strengthened superalloys, refractory metals, or carbon-carbon composites are used, depending on the specific temperatures involved.

Nickel-alloy heat pipes can be used where local temperatures may be as high as 3500F. The heat pipe would transfer the heat to a large, cooler area, from which it can be radiated. The heat pipes operate at about 1600F.

The new system is intended to survive the expected 100-mission lifetime of future Shuttles, and it should be more durable than the fibrous or ceramic tiles, which have to be carefully monitored for surface fraying, erosion, and cracking. The metal panels—which are bayonet-mounted to the primary structure, to allow for expansion and thermal-stress relief, and are isolated from the strains of the primary structure—should also be more resistant to thermal and mechanical stresses and competitive in weight.

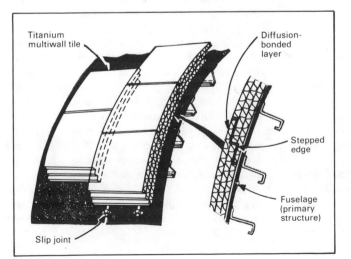

Titanium multiwall tile

Diffusion-bonded layer

Stepped edge

Fuselage (primary structure)

Slip joint

Reprinted from *Modern Plastics*, December 1981

Liftoff is set for advanced thermoplastic composite materials

By A. Stuart Wood

Fiber/engineering-resin prepreg tapes and fabrics now in development show performance advantages over thermoset composites, and they can be turned into structural components in a fraction of the previous time

Space shuttle blasts off from Cape Canaveral. One of the tasks the vehicle is scheduled to perform: pultruding carbon/polysulfone beamstock in orbit. [Photo, NASA]

Advanced thermoplastic composites may become as familiar in this decade as their established thermoset counterparts became in the '70s . . . and their commercial acceptance could be much broader. Because even though the immediate uses for which the thermoplastic materials are being developed—such as truss beams that will actually be pultruded in the Space Shuttle and filament-wound vehicular components—are highly specialized, they are paving the way for the emergence of a whole new and promising technology.

Offering some distinct performance advantages over thermoset composites—notably elongations to break in the 20 to 30% range, against only 0.5 to 7% for thermosets—advanced thermoplastic composites also hold the promise of significantly increased production rates in such processes as pultrusion, filament winding, automated tape winding, and various kinds of stamping and hot forming operations.

This developing technology is being accelerated by material suppliers such as Hercules, Celanese, Du Pont, Union Carbide, Fiberite, U.S. Polymeric, and equipment builders like Goldsworthy Engineering and McClean Anderson. The research and development effort that makes it all practicable is being funded by government agencies and private industry, including the Dept. of Energy, National Aeronautics and Space Administration (NASA), Naval Air Systems Command (NASC), and such firms as Boeing Aerospace Co., and Grumman Aerospace Corp.

Pultruding in space

A current Goldsworthy project involving advanced thermoplastic composites sounds like something out of science fiction . . . but it is an entirely feasible expression of late 20th-century technology. It is a system for the automated production of a composite truss beam by a modified pultrusion process. In development for NASA, the system will be used in the Space Shuttle to supply beamstock for the construction of large orbiting structures.

The system is in fact a blending of two systems: one for producing a complex-geometry carbon/polysulfone ribbon on earth, and another for converting this in orbit into beamstock.

In the first system (see illustrations on the facing page) prepreg carbon tow is pulled from creels through a carding plate and drawn longitudinally onto a stationary mandrel, after which a ring-shaped winding head applies additional carbon fiber, laying it down in a 45-deg. pattern around the longitudinal tow "tube." The wound prepreg tube next passes through an induction heating station that melts the polysulfone at a temperature of around 650°F. The molten tube is then squashed flat and consolidated into ribbon form by continuous opposed belt laminating. The finished ribbon, which is approximately 8 in. wide and 30 mils thick, is taken up on a storage reel.

In the second orbital-processing system, 8-in. ribbon is itself processed through equipment that is described as a "beam builder." This is basically a pultrusion system that makes use of three inductively heated "heat pipes" in order to raise the temperature of three separate ribbons to 600°F. at the fold lines for subsequent forming into open-ended, triangular-shaped corner members of the beam. Forming and guidance tooling are used to accomplish this last task, after which the unclosed members pass through a final heat die that welds lap seams, closing the triangles. From here the members enter a short chill die section where they're radiantly cooled to about 500°F. and solidify into their final structural form.

Downstream of this, corner members are joined to make the completed truss beam by intercostal struts, set at 90- and 45-deg. angles, which are dispensed from cassettes, much as bullets from a clip in the magazine of a gun (see smaller diagram, below right).

Throughput rate of the system is 2 to 6 ft./min., and each of the triangular corner members is about 2½ in. wide. It's reported by the developer stronger than an equivalent aluminum structural member, is 30% lighter, and has zero expansion and contraction in the drastic temperature changes that are encountered in space.

Goldsworthy points out that while the beam builder is an exotic application of advanced thermoplastic composites, it points the way to more mundane uses. The main reason polysulfone won out over candidate thermosets for the truss beam is the former's comparative ease of processing. Thermosets are tough to control, especially in the vacuum and temperatures of space. The simple heat/chill cycle required for forming the polysulfone dispensed with all of the difficult handling, long cure times, and outgassing constraints in space associated with thermosets.

Goldsworthy believes the same rapid and relatively simple processing operation that will take place on board the Space Shuttle can greatly improve the productivity of the composites industry here on earth.

Filament winding polysulfone

One of the first thermoplastic parts to be successfully filament wound is an energy-generating flywheel rotor comprised of wound carbon/polysulfone unidirectional prepreg tape. Developed under the general auspices of the U.S. Department of Energy, and designed and produced by Hercules in cooperation with Sandia Laboratories, the flywheel is intended to serve as a source of "instant energy" for quick starts, passing, and hill climbing in a prototype electric vehicle.

Hercules makes the tape itself, in a technique that involves using a solvent-coating method in order to impregnate its AS4 high-performance continuous carbon fiber with the polysulfone. Unlike a comparable thermoset prepreg, the carbon/polysulfone tape is completely dry, and thus obviously is significantly easy to handle. It is 100 mils wide, 20 mils thick, and about as flexible as a similar steel tape would be, according to its developer.

Total weight of the aluminum-hubbed wheel is 42.4 lb. (38.6 lb. of it wound composite). The part tapers inward from the outer rim from a maximum thickness of 2.88 in. to 1 in. at the hub, and its has an OD of 23.2 in. and an ID of 4.3 in.

To wind it, Hercules used improvised equipment consisting principally of a special tape-delivery head with provision for melting the polysulfone matrix at 700°F. In the process, tape is drawn off a reel and through the heated delivery head, which lays down the molten tape in a programmed pattern under 30-lb. tension. The head also heats the pre-

Two-part pultrusion system makes building blocks in space

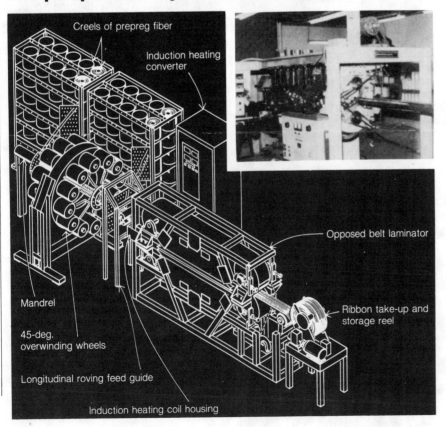

Creels of prepreg fiber
Induction heating converter
Mandrel
45-deg. overwinding wheels
Longitudinal roving feed guide
Induction heating coil housing
Opposed belt laminator
Ribbon take-up and storage reel

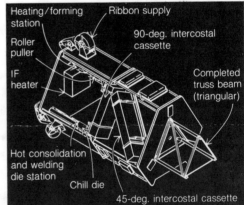

Heating/forming station
Ribbon supply
Roller puller
90-deg. intercostal cassette
IF heater
Completed truss beam (triangular)
Hot consolidation and welding die station
Chill die
45-deg. intercostal cassette

In ribbon-making system (at left and photo), carbon/polysulfone prepreg fiber is drawn over stationary mandrel, overwound, and consolidated into flat ribbon by continuous opposed-belt laminating. In orbital-reprocessing pultrusion system (above), the 8-in.-wide ribbon is formed into three truss-beam members through inductively heated "heat pipes" and chill dies, after which connecting crossmember struts are inserted by intercostal cassettes, much as bullets from the magazine of a gun. [Diagrams and photo, Goldsworthy Engineering]

Thermoplastic prepeg materials like those above can be processed into structural parts without the time-consuming cure cycles needed for thermosets. [Photo, Fiberite]

Energy-generating flywheel for electric car is mostly filament-wound carbon/polysulfone. Weighing 42.4 lb., it's capable of 25,000 r.p.m. [Photo, Hercules]

viously wound tape at the point of lay-down, thereby creating a continuously fused entity.

Hercules says it selected polysulfone as the base material for its good physical properties, as well as for low viscosity for fiber impregnation.

Why a thermoplastic and not a thermoset for the application? According to Hercules, the chief reason was the thermoplastic's vastly superior integrity in the complex construction involved, and in a part that must withstand the stress of spinning at 25,000 r.p.m. (steel flies apart at 20,000 r.p.m.). Prior use of a conventional carbon/epoxy prepreg yielded an uncured winding that was comparatively "loose" or "spongy," and which had a tendency to develop interlayer wrinkles, with consequent fiber misorientation and flaws that are totally unacceptable in a high-performance finished part.

Time of manufacture was another essential consideration—assuming that the part eventually goes into volume production (where productivity increase is important to profitability). In order to minimize fiber misorientation, winding of the epoxy wheel had to be interrupted periodically while the partially wound part was cured. So although winding speed for the polysulfone, which requires no such cure cycles, was much slower

(20 ft./min. versus 300 for the epoxy), overall production time to a finished part is reportedly less than a third that required for the epoxy.

Equipment manufacturers working on the concept believe that winding speed can be considerably improved over the 20 ft./min. obtained by Hercules—with consequent improvement in its commercial acceptance. Goldsworthy Engineering, for example, envisages the practical development of a system involving a heated mandrel and a delivery head equipped with a radio-frequency or inductive heating element. And McClean Anderson is currently developing a delivery system geared to winding thermoplastics for Boeing, which is interested in the attractive cost/performance offered by thermoplastic prepregs in the manufacture of aerospace and aircraft components.

Ongoing material developments

While much of the initial work with advanced thermoplastic composites has been with polysulfone, some of the more recent developments include a number of other high-performance engineering resins—most notably nylon, polyester, polycarbonate, phenylene oxide, polyamide-imide, and the new polyetheretherketone (PEEK).

Using solvent- and hot-melt-coating

techniques, these materials (and other suitable thermoplastics) can be used to impregnate carbon and other high-performance continuous fibers to produce unidirectional tapes or fabrics.

In the first and currently most widely used approach, the resin selected is dissolved in solvent, after which a specially surface-treated fiber is impregnated with the solution by means of a coating head. The solvent is subsequently removed from the fiber by the process of evaporation.

It's this last step, with its unfavorable environmental connotations, that has led to consideration of the second approach: the hot-melt coating alternative. High heat (temperatures of 700 to 800°F.) replaces the solvent in order to reduce the plastic material to a suitable viscosity for thorough fiber impregnation, which is accomplished by means of a specially adapted head.

The coated fiber is then taken up on a heated drum and fused together in parallel to make tapes of various widths, or it's used to make woven fabrics. These prepregs can then be shipped, stored, and processed at potentially attractive savings over traditional thermoset prepregs.

Boeing, for instance, notes that while automation would be desirable in reducing the high cost of producing carbon/epoxy aircraft sections by hand layup methods, the capital outlay entailed in developing the necessary equipment would be formidable. It says that a promising alternative is advanced thermoplastic composites, which it has determined can be sheared, stamped, dimpled, and hot-formed using adapted metalworking equipment. The materials can of course also be wound and otherwise processed as described earlier in this article.

Though Boeing is unwilling to go into details, other manufacturers involved in this fledgling technology confirm that developmental nylon prepregs have been heat-stamped and hot-formed to make high-strength (typically 200 k.s.i. tensile, 20 m.s.i. modulus) finished parts in process cycle times of under 1 min., versus 30 min. to several hours for their thermoset equivalents.

And new resins are being developed as a direct result of this work. For example, as part of its program to research new thermoplastic prepreg materials for structural and other uses in aircraft, Boeing has come up with a super polysulfone dubbed NTS resin. The material's big claim to fame—it's high resistance to aircraft hydraulic fluid and cleaning fluids that are routinely used in normal flight and maintenance operations, and which attack polysulfone and many other thermoplastics. ■

Key properties of carbon/polysulfone prepreg tape				
	Temperature at which measured, °F.			
Property	−67	72	180	250
Tensile strength, k.s.i.	200	188	192	180
Tensile modulus, m.s.i.	18.1	16.3	16.2	17.5
Flexural strength, k.s.i.	230	190	156	135
Flexural modulus, m.s.i.	20.0	17.8	19.1	20.0
Thermal conductivity, Kz. (B.t.u.-ft./ft.^{2}-hr.-°F.)	0.26	0.30	0.34	0.36

Source: Hercules.

Reprinted from *Iron Age*, January 12, 1983

METAL MATRIX COMPOSITES POSE A BIG CHALLENGE TO CONVENTIONAL ALLOYS

A Federal ruling prevents us from saying too much about metal matrix composites. In sticking to the law, we can still discuss the basics.

BY ROBERT R. IRVING

In materials, the big buzz word in recent years has been composites, with particular emphasis on the graphite-epoxy composites. Some say that these unusual materials will, in time, give the traditional metal products a real run for their money.

With all of this attention being focused on the graphite epoxies, another perhaps more spectacular materials technology is gathering steam throughout the world. That technology is metal matrix composites.

The big talk in this country in terms of metal matrix composites are the struts on the Space Shuttle. The only true production use of these materials, the struts are made of boron/aluminum. On the Space Shuttle, the 243 tubular struts of boron/aluminum are used in the mid-fuselage structure (below) of the orbiter vehicle. This development won out over aluminum alloy extrusions.

The importance of metal matrix composites cannot be ignored. The Federal government, for example, came out with a ruling about three years ago that prevents the release of a substantial amount of classified information on the subject to the public. Issued by the Department of State, this ruling falls under the purview of the United States Munitions List, Section 121-01, International Traffic in Arms Regulations.

The ruling states, in part, that such classified information "shall not be exported or released to foreign nationals in the U.S. or abroad without a validated export license. Penalty for violation," the ruling warns, "is up to 2 years imprisonment and a fine of $100,000."

The government means business, so it seems.

In the non-classified sector, there have been two developments of particu-

Ways Metal Matrix Composites Can Be Fabricated

Matrix State	Method Used to Combine Matrix with Fibers	Method Used to Form and Shape Composite	Composite System Matrix (Fiber)
Molten	Infiltration into bundles of parallel fibers	No additional working	Cu (W); Cu-alloy (W); Al (glass); Ag, Al, Ni (Al2O3)
	Infiltration into fiber mats	Rolling and machining	Ag (steel, Mo)
	Fiber solidified directly from melt	Some specimens rolled	Al (NiAl); Ta (Ta2C); Cb (Cb2C)
Powder	Mix	Hot press	Al (steel)
	Blend, cold press, sinter	Hot extrusion, anneal	Ti, Ti-alloy (Mo)
	Mix, press	Hot extrusion swage	Al (W)
	Mix, hot press, cold press, sinter	Hot roll, swage or forge	Ni/Cr, Co-alloy, Co, steel (W)
	Mix, melt matrix, blend	Extrude, sinter, cold roll	Ni/Cr, Fe (Al2O3)
	Blend, dry	Extrude, sinter, roll	Ag, Ni (Si3N4)
Molec-ular	Electroplating	No additional working	Ni (Al2O3, W)
	Electroplating	No additional working	Ni (Al2O3)
	Vapor plating, electrocodeposition, hot press	Hot roll	Ni (Al2O3)
Sheet or foil	Alternating layers of sheet and fibers	Diffusion bond	Al (steel, Be); Al (B, Be)
	Alternating layers of sheet and fibers	Diffusion bond	Ti, Ti-alloy (B); Ti (Be)
	Alternating layers of sheet and wire mesh	Hot roll	Al (stainless steel)
Combin-ation	Draw fibers through molten matrix	Hot press coated fibers	Al (SiO2)
	Electroplate fibers	Encapsulate in can, draw, sinter	Ag (steel)

Star structure was forged from DW Al 20, DWA Composite Specialties' trade name for its discontinuous product.

This cylinder was extruded by DWA Composite Specialities from discontinuous reinforced aluminum powder metal.

lar note, one in this country and the other in Japan.

The American development involves the metal matrix struts on the Space Shuttle. It represents the only true production use of such advanced composites in this country. The end result is a material having the stiffness or strength of steel and the weight of aluminum.

The actual material consists of uncoated tungsten core 5.6-mil diam boron filaments in a 6061-F aluminum alloy matrix. This composite serves as the material for the 243 tubular truss members used in the mid-fuselage structure of the orbiter vehicle. The tubes make up the main frame and rib struts, the frame stabilizing braces and the nose landing gear drag-brace struts.

The composite struts exceed all performance requirements for the orbiter mid-fuselage structure and have saved about 320 lbs over the original concept, which was based on aluminum alloy extrusions. The weight reduction is 44 pct. There is also a savings in space.

Developed by the Convair Division of General Dynamics Corp., San Diego, Calif., the tubes are fabricated from unidirectional, single layer, boron filament-reinforced, diffusion-bonded sheet material. The reinforcing material is oriented along the length of the tubes.

The material is consolidated into a multilayer tube at the same time it is joined to 6A1-4V titanium end collars by diffusion at high temperature and high pressure in an autoclave. The titanium alloy end fittings are later electron beam welded to the collars.

Eighty-nine combinations of boron/aluminum tube length, thickness and diameter are used on the Space Shuttle. The largest is 89.8 in. long, 3.63 in. diam, and weighs 7¼ lbs. The smallest is 23.8 in. long, 1 in. diam and weighs only 3.3 lbs.

The big news from Japan comes from the automotive industry. It is another production application involving the piston in Toyota's new turbo-diesel engine. The top plate and ring groove of this piston are both made from a metal matrix composite. The reinforcement is an alumina-silica material and the matrix is an aluminum casting alloy. This composite is replacing a Ni-Resist cast material.

Toyota officials are claiming excellent wear and seizure resistance for this new piston.

The details surrounding this development are sketchy and shrouded in a certain amount of secrecy. Nevertheless, several facts and opinions have leaked out. One source believes that the reinforcement material is actually an improved modification of Babcock & Wilcox's Kaowool, which is an alumina-silica refractory material. Another source suspects that the composite components are formed into shape by squeeze casting.

"This application at Toyota," says one qualified spokesman, "could break this whole thing about metal matrix composites wide open. Toyota is achieving the properties it wants. The development fits right into the company's existing production scheme and it's cost-effective."

Japan has made composites a national priority. According to the same American spokesman, Japan is at least five, if not 10, years ahead of this country in composite technology. "When it comes to advanced composites," he predicts, "Japan will mop us up."

Another authority claims that France is rapidly becoming strong technologically in metal matrix composites. West Germany and the Soviet Union are also putting a lot of effort into it.

In this country, both the aerospace industry and the military are familiar with metal matrix composites. With the exception of the boron/aluminum struts for the Space Shuttle, however, practically all of the work has been developmental. Automotive is interested, but the run-of-the-mill of industry is not acquainted with the technology at all.

Why is this technology so important?

"A great advantage in composites," says Carl Zweben, advanced technology manager, General Electric Co., Space Systems Division, Valley Forge Space Center, Pennsylvania, "is that you have so many combinations of fibers or reinforcements and matrix materials to look at. You can do a lot of tailoring in order to achieve many unique characteristics that you would never be able to achieve otherwise."

Dr. Zweben also discusses what happens to conventional aluminum when it is in composite form. When the aluminum is reinforced, the resultant composite becomes a different material entirely. Depending on the reinforcement used, it can have up to five times the modulus and strength of unreinforced aluminum.

"Use of an alumina/aluminum composite," he adds, "raises the temperature capability of aluminum just enough to make aluminum work in an engine. It also brings the coefficient of expansion of aluminum close to that of steel. This is a very important point to consider."

Incidentally, word has it that the Japanese are indeed looking at engine blocks made from metal matrix composites.

In this country the prime source of information on the technology is the two-year-old DOD Metal Matrix Composites Information Analysis Center (MMCIAC) in Santa Barbara, Calif. The center is operated by Kaman Tempo.

The MMCIAC provides the facilities and capabilities to do the following:

—Identify, collect, process, store and disseminate authoritative metal matrix composite information.

How the Various Composites Break Down

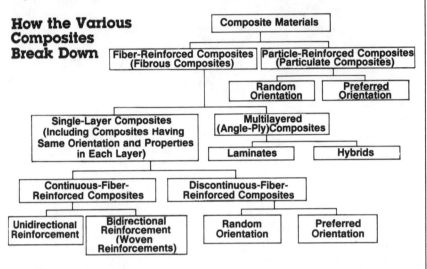

—Prepare or sponsor the preparation of the necessary products and services to communicate this information to researchers, practicing specialists, manufacturers and other users with interests and concerns in the subject.

—Coordinate and augment existing information activities to improve the transmittal of this information to interested organizations and individuals in the government, military and private sector.

For example, the center will be co-sponsoring a short course on metal matrix composites at the University of Maryland from Feb. 26 through March 4.

Also, Jacques E. Schoutens of Kaman Tempo has prepared a tutorial on metal matrix composites. In it, he points out that composites are capable of improving various properties. Among them are strength, stiffness, fatigue life, wear resistance, weight, corrosion resistance, temperature-dependent behavior, thermal insulation, thermal conductivity, acoustical insulation and attractiveness.

Also, he notes that metal matrix composites essentially consist of two materials. One is the bulk material, if you will, known as the matrix. The matrix will often be a low-density metal,

such as aluminum or magnesium. Filler material of some sort is added to the matrix. The matrix then holds these fillers or fibers or reinforcements together to enable the transfer of stresses and loads to the fibers themselves.

The fibers, on the other hand, are the materials that carry the major stresses and loads. Fibers can further be broken down into two categories—continuous and discontinuous fibers. The boron/aluminum composites used for the struts in the Space Shuttle were produced from continuous boron fibers. They have outstanding properties unidirectionally, which is just what is needed for that type of application.

Many experts claim, however, that the big future for metal matrix composites will be with the discontinuous fibers, which are in the form of chopped fibers, whiskers and particles. Composites made from discontinuous fibers lend themselves more readily to such conventional metalworking operations as forging, extrusion, squeeze casting and welding.

Within a composite structure, the fibers will occupy anywhere from 25 to 80 pct of the mass of the structure.

"There is quite a variety of continuous reinforcement materials," points out Dr. Zweben, "and the list keeps growing."

Among the leading candidate continuous fibers are boron, silicon carbide, graphite and alumina. Continuous boron fiber in a matrix of aluminum is, in fact, the most widely used reinforcement to date. To ready this fiber for use in a metal matrix, boron is deposited by chemical vapor deposition onto a tungsten or carbon core that might only be ½ mil in diameter. The core is then built up to a 4 to 8 mil diameter.

Chemical vapor deposition, incidentally, is also used to deposit silicon carbide onto carbon cores.

According to Dr. Zweben, who is also

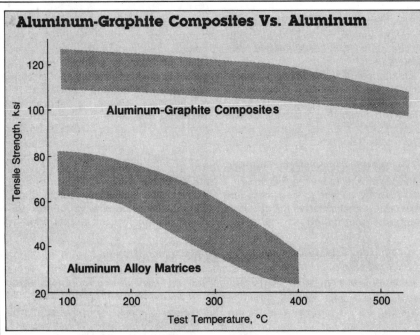

Aluminum-Graphite Composites Vs. Aluminum

Aluminum-Graphite Composites

Aluminum Alloy Matrices

Tensile Strength, ksi

Test Temperature, °C

a technical consultant to MMCIAC, a large number of graphite or carbon fibers are on the market. The precursor materials for these fibers are polyacrilonitrile (PAN) and pitch.

Graphite fibers can be divided somewhat arbitrarily, he says, into three categories: high strength, high modulus and ultra-high modulus. At present, much of the work with graphite fibers is focused on high modulus and ultra-high modulus pitch fibers in aluminum and magnesium matrices.

Some sort of reaction barrier is often needed to prevent any harmful chemical reaction between the fiber and matrix materials. Boron fibers, for instance, can react with matrix metals at high temperatures. Several barrier coatings have been developed for such conditions. Included are silicon carbide, boron carbide and boron nitride.

Then there are Borsic fibers produced by vapor depositing boron onto fine tungsten wire, then coating that product with silicon carbide to provide the reaction barrier needed between the boron and the matrix material. Borsic fibers are made by Composite Technology, Inc., Bound Brook, Conn.

The main whisker and particle reinforcements to date are silicon carbide, and they have been used in aluminum alloy matrices. There is also interest in aluminum and magnesium reinforced with alumina whiskers.

Metal wires have also been developed as reinforcing materials for high temperature applications, such as jet engine components. Examples here are wires of tungsten and molybdenum used to reinforce such matrix materials as titanium and certain superalloys.

Both boron and graphite were selling for several hundred dollars per pound about 10 years ago. Graphite prices have since been reduced and are now moving toward a plateau of about $20 per lb or less. Boron, on the other hand, is only coming down toward the $100 per lb mark.

One reason behind the relatively high price of boron is the present manufacturing process of depositing the boron onto a tungsten core or substrate. If the boron were to be deposited onto a glass rather than a tungsten substrate, further economies could be realized. If such a substitution can be made, there is also an advantage in that smaller diameter fibers could be produced from glass.

Although the very idea of paying $300 per lb for a material is enough to make the typical automotive engineer break out into fits of laughter, this high basic cost for a fiber can be very misleading.

In his tutorial, Mr. Schoutens looks at cost from the near-net-shape viewpoint and cites, as an example, the fabrication of spars and longerons for modern aircraft, to prove his point.

"Spars and longerons in aircraft wings," he explains, "are beam elements that are usually tapered in depth and width, with holes in their webs to reduce weight.

"The fabrication of spars and longerons from aluminum or other aircraft-type lightweight alloys consists of machining them from large blanks of materials that may weigh as much as seven times the final weight. This results in 600 pct scrap. Spars have been fabricated from composite materials with as little as 10 pct scrap," he adds.

One company that has done a great deal of work in the fabrication of components from metal matrix components

is DWA Composite Specialities, Inc., Chatsworth, Calif. One of its recent jobs is to make the boom waveguides for the high gain antenna on NASA's space telescope. The prime contractor on this project is Lockheed Missiles & Space Co.

DWA is making the boom waveguides from graphite-aluminum. Graphite was chosen for this application because of its high stiffness-to-density ratio and its low coefficient of expansion in the fiber direction. Once the space telescope is completed, it will be a permanent national astronomical observatory that will orbit 310 miles above the earth.

The California company works with a wide range of materials. The fibers include most of those already mentioned, a number of proprietary grades and such materials as boron nitride, boron carbide, titanium nitride and titanium carbide. Matrix materials include 6061 and 2024 aluminum, 6A1-4V titanium, and some magnesium and copper.

Looking toward the future, Joseph F. Dolowy, Jr., president of DWA Composite Specialities, is high on the potential for discontinuous reinforced composites. He goes as far as to say that he sees a million pound per year market developing for these materials within a few years.

Mr. Dolowy also sees more involvement on the part of basic metals companies in the production of these materials.

"Discontinuous reinforced composites provide a stress/strain performance similar to those of metals," he says, "but the composite material is isotropic and provides high stiffness. Like metals, these composites can be rolled, forged and extruded. The properties of these materials are great and they are not as expensive as continuous reinforcement materials."

Mr. Dolowy claims that both the commercial aircraft builders and the automotive manufacturers are interested in discontinuous reinforced composites. Silicon carbide whiskers are being evaluated, for example, as fillers for aircraft wings.

The technology, however, has been experiencing some growing pains. Lack of ductility has been one problem, although DWA has succeeded in improving that situation. Another problem is joining. The basic properties of one of these materials are one thing, but what are the properties in the as-fastened, as-riveted or as-welded condition? They are not as good, that's for sure. They never are, even in metals.

Still, DWA has fastened, riveted, resistance welded, and Tig and Mig welded discontinuous reinforced composites successfully.

Spurred on by the interest in discontinuous fibers, DWA Composite Specialties is working on silicon carbide particles in aluminum, while ARCO Metals in Greer, S.C., is working on the development of silicon carbide whiskers in aluminum.

Among the newer developments in fiber materials, there is SCS from Avco Specialty Materials Division, Lowell, Mass. A silicon carbide, it is being touted as a lower cost replacement for boron. Its properties are said to be similar to boron and it is also easier, according to Avco, to protect against reaction.

Avco also has a silicon carbide monofilament which it claims will sell for $50 per lb someday. Nippon Carbon Co., Ltd. of Japan has developed a multifilament yarn of silicon carbide that also has good long-range cost potential. It is available in this country through Dow Corning, Midland, Mich.

Another reinforcement material of interest is alumina which retains its properties at elevated temperatures and does not react with the matrix material. The Du Pont Co.'s FP material is a good example of a commercially available alumina fiber.

Elsewhere, graphite-reinforced copper and aluminum composites are of interest for high strength electrical conductors. As far as bearing materials are concerned, there are graphite/aluminum, graphite/lead and graphite/zinc composites.

How do the strengths and stiffnesses of various composite materials compare with one another? How do they compare with bulk metals? Consider fibers of high strength graphite, boron, high modulus graphite and beryllium and consider these same fibers as they perform both in unidirectional and biaxially isotropic conditions. Then, compare these advanced composites with a group of bulk metals, such as steel, titanium, aluminum and beryllium.

According to the DOD Metal Matrix Composites Information Center, the commonly accepted way to express the effectiveness of strength or stiffness of a material is as a ratio of stiffness or strength to density.

Boron fibers exhibit high stiffness and strength efficiencies. When used in a lamina as unidirectional fibers, the relative strength of boron drops significantly while the relative stiffness drops only slightly.

In a biaxially isotropic configuration, boron/epoxy is still stiffer than steel or titanium, although it is of the same relative strength. High strength graphite fibers and composites exhibit similar behavior. However, high modulus graphite fibers, in which stiffness is greater in all configurations than the other materials, have generally lower relative strengths (even lower than aluminum when used in a biaxially isotropic configuration).

S-glass/epoxy in a unidirectional lay-up has about 2.5 times the relative strength of steel or titanium. Yet, it is less stiff than either metal in a biaxially isotropic configuration.

Beryllium, on the other hand, has about six times the relative stiffness of steel, titanium or aluminum, but it has no greater strength. Beryllium wires are much stronger, but no stiffer than bulk beryllium. In a matrix, beryllium exhibits some of the same general traits as do other composite materials.

Advanced composites are not only new materials, they are also a new engineering language.

In his tutorial, Mr. Schoutens points out that composites have many characteristics that are different from conventional engineering materials. He notes that common engineering materials are homogeneous and isotropic, and defines both terms.

"A homogeneous body has properties that are uniform in every direction and, consequently, are not a function of position in the body. An isotropic body," he says, "has material properties that are the same in every direction at a point in the body. This means that the properties are not a function or orientation at a point in the body."

In contrast, he further notes, composite materials are often inhomogeneous and nonisotropic (orthotropic or anisotropic).

Continuing, Mr. Schoutens says that an inhomogeneous body has nonuniform properties, that is, the properties are a function of the position in the body.

An orthotropic body, he notes, has the same properties in every direction. "Materials reinforced with continuous fibers, however, are not isotropic," he explains. "Stiffness and strength parallel to the fibers are much greater than in other directions."

Finally, an anisotropic body has material properties that are different in all directions at a point in the body. There are no planes of material property symmetry, and the properties are a function of orientation at a point in the body.

"Because of the inherent inhomogeneous nature of composite materials," explains Mr. Schoutens, "they are usually studied from two different approaches—micromechanics and macromechanics."

Micromechanics, he states, is the study of composite material behavior wherein the interaction of the constituent materials is examined on the microscopic scale.

On the other hand, macromechanics is the study of composite material behavior wherein the material is assumed to be homogeneous and the effects of the constituent materials are detected only as averaged apparent properties of the composite.

The concepts of micromechanics and macromechanics are used to design composite materials to meet particular structural requirements.

The compilation of property data on metal matrix composites has been something of a problem, observes Dr. Zweben of General Electric.

"Although metal matrix composites tend to get lumped into general categories, such as boron/aluminum," he says, "the materials falling into these classifications can differ significantly from one another, depending on the exact type and source of fiber and matrix used and the method of fabrication and processing history."

To prove his point, Dr. Zweben notes that differences of 46 pct in longitudinal compressive strength and 110 pct in interlaminar shear strength were found in the reported properties of six boron/aluminum composites selected at random.

"One of the major concerns in the field of advanced composites," he points out, "is that, at the present time, property data are relatively sparse compared to those available for structural metals and polymer matrix composites, with the exception of boron/aluminum. This is particularly true for multidirectional laminates."

Mr. Schoutens says that the Department of Defense has maintained a strong interest in promoting metal matrix composite materials over the past 15 years. The DOD's investment in this technology over that time span has been between $70 million and $80 million. The government's interest has intensified even more lately.

Metal matrix composites also appear to be destined to play a very important role in the area of replacement and substitution of critical materials. It's quite feasible that they may be replacing the likes of chromium-based and cobalt-based alloys, both of which are high on the critical list.

If they do, they will have entirely different compositions than the materials they would be replacing.

Dr. Zweben says: "Metal matrix composites can be used wherever properties and performance are important." □

Reprinted from *SAMPE Journal*, May/June 1984

BEST PAPER AWARD — 15TH NATIONAL SAMPE TECHNICAL CONFERENCE — OCTOBER 1983

APPLICATION OF BERYLLIUM ON THE SPACE SHUTTLE ORBITER

L.B. Norwood

*Space Transportation and Systems Group,
Rockwell International, Downey, California*

ABSTRACT

This paper identifies the four different types of beryllium materials used in the various component applications on the Space Shuttle orbiter and discusses why each was selected. The applications consist of six crew module windshield spacers, six inner crew module windshield retainers, four crew module windshield beams, one navigation base with beryllium edge members and facings, two large external tank (ET) umbilical doors, two star tracker adapter plates, and two heat sinks in the forward windshield area. Altogether there are 22 mechanically attached machine beryllium parts on the Shuttle, not including the adhesive-bonded navigation base, which has five edge member fittings, two facings, and eight splices. This is the first time that hot-pressed beryllium has been successfully used on a pressurized manned aerospace vehicle. Beryllium was used in the crew module windshield, navigation base, and ET door areas primarily because of its high stiffness to weight. This resulted in a substantial weight savings to the orbiter. However, because of beryllium's limited ductility and notch sensitive nature, the real challenge lay in producing critical parts free of "twinning," i.e., stress-free machined surfaces.

INTRODUCTION

When the Shuttle orbiter design phase began, in 1973 and 1974, there was no plan to use beryllium in either the structure or system components. As the design of the horizontal flight test orbiter, the *Enterprise* (OV-101), was completed and the orbiter was being

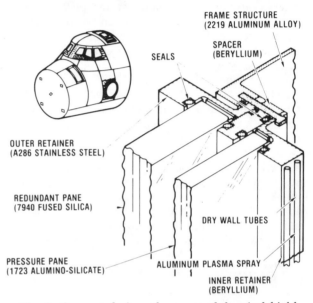

Fig. 2. Current design of crew module windshield.

fabricated, an analysis indicated that the orbiter was overweight. This triggered a series of weight reduction studies, which included consideration of beryllium in areas where its high stiffness to density, in particular, could be utilized.

The first weight trade study to recommend beryllium was completed and approved in March 1975. It recommended replacement of six crew module windshield 2124 Al alloy spacers and six A286 iron alloy inner retainers, as well as four external A286 windshields beams, with parts produced from hot-pressed beryllium block. The design concept is shown in Figs. 1 and 2. It was estimated that the weight saved was 128 pounds at a cost of $2,268 per pound.

In the fall of 1975 a series of inertial guidance system design studies resulted in a decision to design and fabricate an adhesive-bonded honeycomb sandwich structure with beryllium edge members and facings to produce a high stiffness-to-weight ratio, thermally stable navigation base for *Columbia*. These studies consisted of defining the critical design constraints, structural design requirements, ground supprt equipment (GSE), alignment, and mounting techniques. This was followed by a complete stress and environmental analysis and qualification test plan. Results indicated that a bonded, beryllium, tapered, honeycomb box structure was the most feasible approach to satisfy the critical engineering requirements for the stable

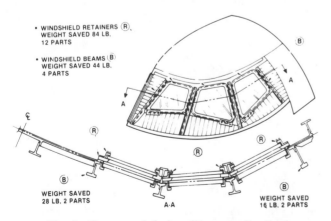

Fig. 1. Crew module beryllium applications.

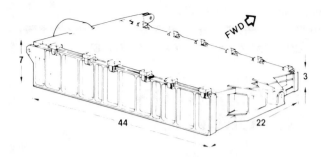

Fig. 3. Adhesive bonded navigation base assembly.

navigation base. The base is required for mounting three inertial measurement units and two electro-optical star trackers as sensors for navigation of the Shuttle orbiter in earth orbit. An aluminum star tracker boom is attached to the navigation base, and two beryllium adapters are attached between the boom and two star trackers. The navigation base design concept is shown in Fig. 3, the star tracker boom in Fig. 4, and the installation concept in Figs. 5 and 6.

In May, 1976, a third design study was completed on the two external tank (ET) umbilical doors. This study compared a double-skin aluminum door concept, two different aluminum waffle door concepts, and a tapered-plate beryllium door concept. After an engineering stress, thermal, and operation design analysis was completed, the tapered-plate beryllium door was selected. Again, beryllium was selected primarily because it had the lowest weight for the required stiffness. The overall cost increase per pound of weight saved was estimated to be only $1,350/lb.

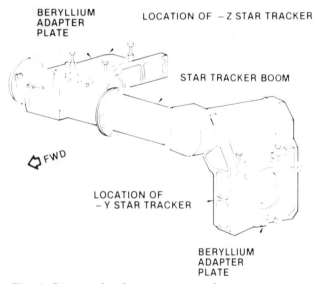

Fig. 4. Star tracker boom portion of navigation base.

DESIGN REQUIREMENTS AND CONSIDERATIONS

All of the previously mentioned structural designs needed beryllium to satisfy the high stiffness-to-density requirement and size limitations even though beryllium has serious drawbacks. It was well known that beryllium was expensive, had limited ductility, was notch sensitive, and was difficult to machine without surface damage. It also had a toxicity hazard. Therefore, a great number of design considerations were made and ground rules established to ensure success. They included:

1. Select the most ductile types of beryllium for structural applications. A minimum of 3-percent elongation in all directions is required. Each hot-pressed block is qualified by testing a production coupon accompanying each part.
2. Use conservative design allowables, resulting in f/s = 2.0 in tension and 1.5 in compression.
3. Require acid etch stress relief for all machined parts to obtain twin-free surfaces.
4. Prohibit metal removal on assembly or installation.
5. Specify a 125-microinch or better surface finish.
6. Avoid tapped holes and inserts on structurally loaded applications.
7. Avoid pressed-in bushings or inserts.
8. Ensure that all holes are match drilled at the detail level.
9. Anodize where feasible or provide adequate corrosion protection finishes between dissimilar materials.
10. Specify liberal filled radii on structural members.
11. Assemble by mechanical fastening or adhesive bonding.

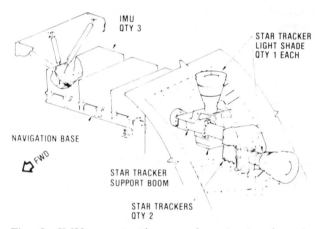
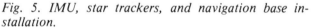

Fig. 5. IMU, star trackers, and navigation base installation.

Windshield Retainer and Post Beam Design Requirements

Since the windshield frame structure had already been designed and fabricated for the *Enterprise,* the beryllium retainer and spacer had to be designed to allow each new windshield assembly to be interchangeable with the original windshield assemblies. Also, there was to be no loss in vision or change to glass trim. The six inner beryllium retainers were required to have 304L

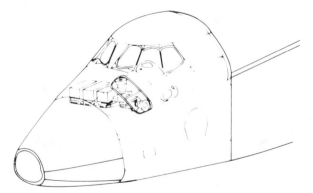

Fig. 6. Location of IMU's and star trackers in crew compartment.

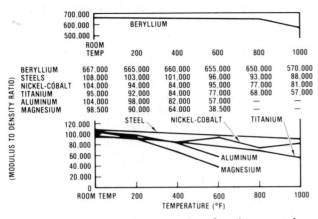

	ROOM TEMP	200	400	600	800	1000
BERYLLIUM	667.000	665.000	660.000	655.000	650.000	570.000
STEELS	108.000	103.000	101.000	96.000	93.000	88.000
NICKEL-COBALT	104.000	94.000	84.000	95.000	77.000	81.000
TITANIUM	95.000	92.000	84.000	77.000	68.000	57.000
ALUMINUM	104.000	98.000	82.000	57.000	—	—
MAGNESIUM	98.500	90.000	64.000	38.500	—	—

Fig. 7. Modulus-to-density ratio of various metals.

CRES tubes attached by aluminum plasma spray to provide thermal conductivity to eliminate condensation (Fig. 2).

Interchangeability of the beryllium windshield post beams with the A286 required that bearing strength allowables be determined for edge distance to hold diameter ratios of less than 1.5 times. Crack growth rate also had to be established. The most important design requirement was that the beams and windshield retainer assemblies be stiff enough to prevent excessive deflection. Large deflections could have put bending loads into the glass panes in space when the crew module windshields have a maximum pressure differential of up to 16 psi.

Navigation Base Design Requirements

A primary design requirement was that the navigation base be as stiff and dimensionally stable as possible because of the critical alignments needed for accurate navigation in space. This meant that the navigation base had to be stable in an environment of inertial, vibro-acoustic, and thermal conditions. Also, the location available for installation limited the thickness of the base. Using beryllium, with its high modulus of elasticity and its excellent thermal properties, along with a tapered honeycomb sandwich platform, avoided any avionic impact risk. Satisfaction of the thermal conductivity requirements required that all adhesive bond thicknesses be 0.005 inches or less. In addition, each inertial measurement unit (IMU) to be attached to the navigation base was to be thermally isolated from the base (the IMU's are air-cooled). This required that 6AL-4V titanium isolators and 304L CRES pads be assembled to the navigation base (Fig. 3).

In order that the avionic alignment requirements could be achieved, each IMU four-pad set was required to be machined within ± 500 arc-seconds maximum per axis with respect to each other pad set. For each of the three IMU stations, all four-pad alignment surfaces were required to be machined flat and parallel within 0.0005 inches. (After navigation base assembly, IMU and star tracker pad sets are measured optically to support vehicle navigation and payload pointing.)

The star tracker boom assembly, which attaches to the navigation base, has two mechanically fastened beryllium adapter plates with precision flat surfaces for mounting the two star trackers (Fig. 4). Beryllium was chosen for its high modulus and thermal stability. The

critical alignment surface was required to be nickel plated and lapped within 0.000044 inches over three coplanar surfaces.

External Tank Umbilical Door Design Requirements

The ET umbilical door design requirements are unique because those doors are the only doors that must be open during launch and closed prior to reentry. During ascent, there is a nominal clearance of only 0.80 inches between the tiles on the doors and those on the lower surface of the aft fuselage. This means that the doors must be very stiff to withstand the dynamic and static deflection loads as well as the vibro-acoustic and thermal conditions during launch. With use of a simple beryllium plate, machined from hot-pressed block 1.29 by 50 by 50.25 inches thick, an analysis revealed that the 0.80-inch deflection clearance could be met. Beryllium provided other advantages, too. Its high heat resistance, high specific heat, good thermal conductivity, and low thermal growth ensured that there was no need for internal insulation during launch.

The only critical requirement for fabrication was to maintain the external and internal attachment surfaces flat and parallel within 0.010 inches with a surface finish of 125 microinches or better after etching.

SUITABILITY OF BERYLLIUM

In beryllium applications on the Shuttle orbiter, the designers took advantage of its unique combination of properties. A look at the modulus-to-density comparison in Fig. 7 shows that beryllium is 6.4 times greater than aluminum at room temperture. This advantage was

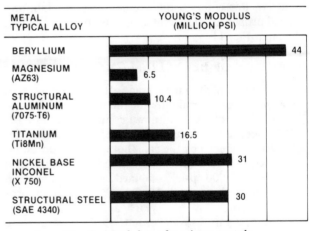

Fig. 8. Modulus of various metals.

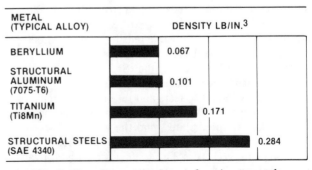

Fig. 9. Density comparison of various metals.

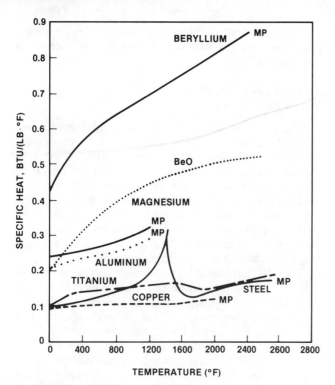

Fig. 10. Specific heat of beryllium compared to other metals.

significant in the design of the navigation base. At 500°F, considered acceptable for non-insulated beryllium ET doors and comparable to insulated aluminum doors at 350°F, the modulus-to-density ratio advantage of beryllium is 7.7 times that of aluminum. In addition, beryllium's high strength-to-density ratio, its high specific heat, and high thermal conductivity proved to be the best choice for the orbiter applications discussed. For comparison, the densities, moduli, and thermal properties of several materials are shown in Fig. 8 through Fig. 11.

SELECTION OF BERYLLIUM GRADES

A premium high purity grade of isostatic pressed block was selected originally for the windshield beams. This beryllium block was known commercially as CIP-HIP 1, a grade supplied by Kawecki Berylco Industries (KBI) and produced by the cold/hot isostatic consolidation process. It had the highest minimum tensile yield properties, 35,000 psi, combined with a minimum elongation of 3 percent. However, KBI ceased producing beryllium a few years ago, leaving Brush Wellman Inc., as its sole producer. As a result, the beams for the remaining two orbiters were made from the S65 premium structural standard purity grade, which has a minimum tensile yield strength of 30,000 psi and a minimum elongation of 3 percent. Since a conservative design allowable tensile yield of 28,000 psi had been used, no design change was made. Also, by this time, confidence in the S65 grade had been established since it had been specified and used on three ship sets of window spacers and retainers, as well as for all of the edge member fittings for the navigation base, star tracker adapter plates, and the two ET doors. Typical Fty values taken from certification test specimens ranged from 33.3 to 36.1 ksi with elongation ranging from 3.4 to 5.9 percent, with the transverse

properties being slightly higher than the longitudinal values, particularly in elongation. The other grades used in the Shuttle were the precision flatness, cross-rolled SR-200E beryllium sheets, used for navigation base face sheets and splices, and standard grade S-200E beryllium block, for windshield heat sinks. The 0.090-inch face sheets had a required minimum Tys of 50 ksi and 10 percent elongation, which was readily obtained. Standard grade block was considered entirely adequate for the heat sinks, since they needed only a high specific heat.

TOOLING

The tooling used for all matched hole drilling, boring, and reaming in detail parts was conventional steel drill fixtures produced from master drill plates or interface control tools. All other metal removal operations were accomplished by programmed numerical control (NC) tape milling machines. All NC tapes were proofed with aluminum prior to machining beryllium. All cutting tools were either solid carbide or had carbide inserts. The adhesive bond fixtures for the navigation base were especially designed for accurately positioning all parts and maintaining that accuracy at bond temperatures up to 250°F with autoclave pressures up to 50 psi.

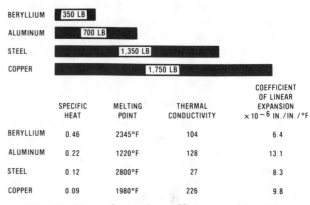

	SPECIFIC HEAT	MELTING POINT	THERMAL CONDUCTIVITY	COEFFICIENT OF LINEAR EXPANSION × 10⁻⁶ IN./IN./°F
BERYLLIUM	0.46	2345°F	104	6.4
ALUMINUM	0.22	1220°F	128	13.1
STEEL	0.12	2800°F	27	8.3
COPPER	0.09	1980°F	226	9.8

Fig. 11. Heat absorption efficiency of beryllium.

Achieving this accuracy meant that the material had to have almost the same coefficient of expansion. Therefore, PH 17-4 steel was selected. The finished bond fixture is shown in Figs. 12 and 13. Figure 12 shows the navigation base detail parts layed up in the bond fixture ready to close out with the adjustable edge member and upper caul plate. Figure 13 shows everything ready for vacuum bagging.

MACHINING

Early in the process development phase of fabricating beryllium parts (from literature surveys and prior experience), it was recognized that beryllium surfaces, to be successful in structural applications on the Shuttle orbiter, had to have all evidence of machine damage eliminated. It had long been known that beryllium, because of its hexagonal close-packed atomic structure, was extremely susceptible to surface damage from machining. This damage is manifested in the form of "twinning" of the crystallographic grain structure to varying depths in the beryllium surface. Twinning can ultimately lead to surface microcracks at intersecting twins, then later travel along the matrix-twin surface until complete fracture occurs. Historically, premature failure of beryllium has been observed to have been initiated at twinned surfaces.

Machining and Etching Requirements

To preclude the possibility of premature failure, the Shuttle Orditer Division of Rockwell International imposed stringent requirements for the machining of the beryllium components for all Space Shuttle applications. The most stringent was that, after machining and etching (to remove twinning), no twinning was to be present in metallographically prepared in-process control specimens. To satisfy this requirement, it was first necessary to qualify the metal removal and acid etch process used by each fabricator and be approved prior to fabrication of production parts. This required documented evidence that all processes and all parts were in compliance with all Shuttle Orbiter Division Engineering requirements. For further control of the metal removal process, it was required that in-process control specimens be fabricated from the same material at the same time as the production parts and be metallographically examined for verification of removal of twinning. The amount of acid etch metal removal finally required to eliminate twinning is now a minimum of 0.006 inches. This was not determined overnight. Prior industry practice was typically 0.003 to 0.004 inches with 0.005 inches considered maximum. The studies conducted at the Autonetics Division of Rockwell International covered a range of 0.005 to 0.010 inches of acid etching following the various machining parameters of (1) depth of cut, (2) cutter speed, and (3) feed rate. When the machining parameters were established on unetched specimens with twinning depths less than 0.006 inches, it was felt that a 0.006-in. etch depth would eliminate all evidence of twinning. This was not the case. Even when one-half of the same specimen was etched, twinning of 0.001 to 0.0015 inches was still present. After an increase up to a 0.010-in. etch with about the same small amount of twin depth remaining, the metallographic polishing tech-

Fig. 12. Navigation base detail parts layed up in bond tool.

Fig. 13. Navigation tool ready for vacuum bagging.

nique became suspect. Indeed, changing the specimen polishing method to reduce surface stress resulted in an end to the problem, and final results indicated a 0.006-in. etch completely removed twinned grains without creating excessive dimensional tolerances (except for holes in thick sections).

Machining Procedure

After a field survey for an experienced beryllium machine source with etching capability the Speed-ring Division of Schiller Industries was selected to fabricate the 12 windshield spacers and retainers. The Autonetics Division of Rockwell International was selected to fabricate all other beryllium detail parts. Both were, therefore, required to be qualified to the Shuttle Orbiter Division's machining specification of Rockwell's Space Transportation and Systems Group. Initial attempts to qualify were unsuccessful, but, finally, both were qualified. This was accomplished by first improving the machining procedure and finally by improving the specimen polishing procedure. While the detail procedure varied slightly between the two sources, all test specimens were at last free of twinning.

Generally for milling operations the cut depths were 0.025 to 0.030 inches for roughing, 0.010 inches for semi-finishing, and 0.005 inches for finishing so the machine damage from each pass would be removed by each succeeding pass. In most cases it was found desirable to make two semi-finishing passes, followed by only one finish pass. On occasion, however, to maintain tolerances, only one semi-finishing pass of 0.010 inches was made, followed by two 0.005-in. finish passes.

Rough cuts were milled at carbide cutter speeds of 157 to 357 SFM at a feed rate of approximately 0.003 IPT. Finish cuts to minimize surface roughness and maintain critical dimensions required speeds as low as 131 SFM at a feed of 0.00075 IPT. Hole drilling, boring, and reaming used speeds ranging from 50 to 80 SFM at feed rates of 0.0005 to 0.00083 IPT.

FABRICATION OF BERYLLIUM STRUCTURES

External Tank Umbilical Doors

The ET umbilical doors were machined from plates of S65 beryllium block cut from a 72-in.-diameter hot pressing produced by Brush-Welman Inc. The as-received slabs from Brush-Welman are 1.190 plus 0.060 minus 0.000 inches thick by 50.00 plus 0.130 minus 0.000 inches by 50.52 plus 0.130 minus 0.000 inches and are required to be flat within 0.030 inches TIR and have a 63-microinch finish. The finished machined and etched doors are 1.100 inches thick in the center tapering to 0.507 inches toward the edges; the entire edge perimeter is 0.900 inches thick. The outer surface is machined flat within 0.010 TIR. The finished width is 49.230 inches, and the length is 49.353 inches. A row of holes is drilled, reamed, and etched in the 0.900-in. thick periphery of each door. Etching 0.006 inches (0.012-in. diameter) from all hole surfaces presented a concern since a "bellmouth" condition occurs in thick

Fig. 15. Umbilical doors installed on Columbia.

accepted by design through MR action. Another problem arose when they were to be assembled: one hole was undersized. However, this was readily corrected by manually swab acid etching and repriming.

Windshield Retainers and Beams

Once all of the tooling for the match holes was tool-proofed and the machining and etching parameters were established, there were few fabrication problems. The etching of the 0.700-in.-thick beryllium windshield spacers caused bellmouthing of the holes of diametrical difference of almost 0.002 inches. There was concern that there might be enough movement in the bolted retainer assembly to adversely affect the bending stiffness of the assembly. The 0.002-in. bellmouthing was accepted, however, after evaluation of the condition and tests on windshield test articles.

The only other problem was attachment of drywall stainless steel tubes to the beryllium inner retainer by aluminum plasma spray. This process involved considerable development and testing. The aluminum plasma spray was used to obtain good heat transfer from the warm water to prevent condensation during the cold environment of space. A test program was established to verify the design and process, as well as to qualify Speedring Manufacturing. The results were successful. The thermal shock test was twice the temperature range expected in flight. The test consisted of three cycles of heating the specimens to 216°F and then quenching into an alcohol and ice bath at 10°F. To date, these plasma-sprayed tubes to beryllium wind-

Fig. 14. External tank umbilical doors.

sections because of non-uniform etch rates. Tests produced a bellmouth diametrical difference of only 0.0016 inches. This was within the ±0.002-in. design tolerance, so it caused no problem. The completed pair of doors is shown in Fig. 14. The doors are corrosion-protected with a conversion coat and primer ready to be assembled. Figure 15 shows the doors installed on the underside of the aft fuselage of the *Columbia* shortly before rollout at the Palmdale (California) facility. Note that the doors are in the latched-open ascent position.

The first two ship sets of doors were fabricated without defects or problems, but the third set presented a slight out-of-flat warpage condition. One was discovered to be 0.020 inches convex, while the other was 0.052 inches concave during machining. Fortunately, the one 0.052 inches out of flat had not been final-machined, and 0.025 inches remained prior to etching. By turning the door over and taking a light cut and then turning it over a second time to complete the finish machining, both doors were corrected and were

Fig. 16. Plasma-sprayed tubes assembled to windshield.

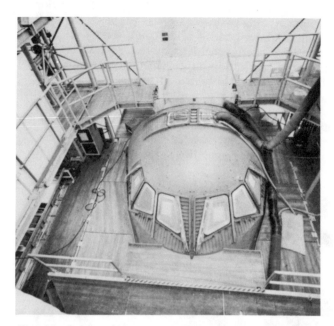

Fig. 17. Orbiter 103 crew module (October 18, 1982).

honeycomb core impregnated with a heat-resistant phenolic resin. The honeycomb core assembly was supplied by Hexcel, which included splicing and precision machining to thickness tolerances of ±0.005

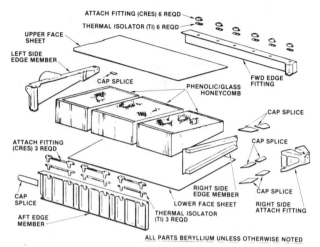

Fig. 18. Navigation base detail parts (exploded view).

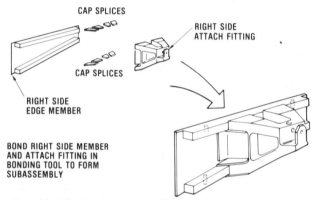

Fig. 19. Navigation base adhesive bond subassembly.

shield retainers have been assembled to three ship sets of double glass pane windshields (Fig. 16).

All six windshield assemblies and the four beryllium beams are shown in Fig. 17 after being installed in the crew module of the third orbiter.

Navigation Base

Of all the beryllium applications, fabrication of the adhesive bonded honeycomb sandwich navigation base is by far the most complex and precise. Figure 18 is an exploded view of the 36 detail parts that are subsequently bonded in six stages. There are five different adhesives used, including the foaming type (FM37) used to splice the HFT 1/8-5.5 glass fabric

Table 1. Multi-Stage Bonding of Navigation Base

Stage	Adhesive	Type of Joint	Cure Temp. (°F)	Time	Pressure (psi)
1. Subassembly	Epoxy-phenolic HT 424	Beryllium splices to be fitting	340 ±10	60-90 min.	10 (min.)
2. Primary	Nitrile epoxy FM 123 & FM 37	Beryllium honeycomb panel	250 ±10	60-90 min.	45 ±5
3. Secondary	FM 123	3 beryllium splices to be panel	190 ±10	6-7 hr.	10-15
4. Secondary	Epoxy No. 206	1 beryllium splice to be edge member	RT or 180 ±10	72 hr. 1-2 hr.	1-10 1-10
5. Fill voids	EA 934	Small tool holes and gaps	RT or 115 to 180	24 hr. 6 hr.	Contact Contact
6. Secondary	Epoxy No. 206	Titanium and CRES fittings to be upper surface	RT or 180 ±10	72 hr. 1-2 hr.	1-10 1-10

Fig. 20. Honeycomb sandwich panel bond assembly.

inches between top and bottom surfaces as well as step cuts. All of the beryllium details were machined, etched, and primed with BR127 primer by the Autonetics Division. The Shuttle Orbiter Division performed all of the assembly bonding oeprations. The six stage bond cycles are described in Table 1. The subassembly bonded in the first stage is shown in Fig. 19. The primary honeycomb sandwich panel bond assembly is presented in Fig. 20. The mechanical property test requirements are shown in Table 2. Fig. 21 shows the navigation base completely bonded, and Fig. 22 shows the base with IMU guide rails attached and painted, ready for assembly of the star tracker boom to it for bench check alignment.

Table 2. Room Temperature Mechanical Property Test Requirements

Adhesive	Type/Test	Minimum Strength Required
HT 424	Lap shear	2,500 psi
FM 123	Lap shear	3,500 psi
FM 123	Honeycomb peel	55 in.-lb.
EA 934	Hardness	75 Shore D
Epoxy No. 206 Grade A Grade A Grade B	Lap shear 90° peel Hardness	3,500 psi 40 lb./in. 70 Shore

CONCLUSIONS

All of the beryllium applications discussed met or exceeded the design requirements as demonstrated by the success of the five earth orbital flights of *Columbia* and the two flights of *Challenger*. The beryllium windshield applications were the most critical structurally since they prevented any excessive deflection of the glass windshield panes when pressurized. This is the first successful application of hot-pressed beryllium on a

pressurized manned spacecraft. From a fabrication viewpoint, the navigation base was probably the most difficult adhesive-bonded honeycomb sandwich structures ever undertaken. It required several tooling, design, and fabrication improvements before a quality stable platform was attained. It was also learned that damaged areas of beryllium can be reworked in most cases. To date, after fabricating over three ship sets of beryllium details and assemblies, only one detail has been scrapped. And it resulted from improper clamping druing a machining operation, which caused the part to fracture.

Fig. 21. Navigation base completely bonded.

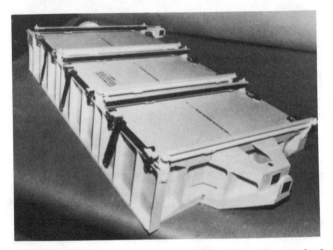

Fig. 22. Navigation base with IMU guide rails attached and painted.

REFERENCES

1. F.E. Smith, M.E. Campbell, T.J. Blucker, C.E. Manry, and Saulietis, *Shuttle Orbiter Stellar-Inertial Reference System.* Space Transportation and Systems Group, Rockwell International, STS 82-0411 (August 1982).
2. D.R. Floyd, W.W. Leslie, D.V. Miley, and R.W. Nokes, *The development of All-Beryllium Riveted Structures.* RFP-2299. Atomics International Division, Rocky Flats Plant, Rockwell International (April 20, 1976).
3. D.B. King, *Fabrication, Material Characterization and Machining Process Evaluation of S-65 Grade Hot-Pressed Beryllium.* AFML-TR-76-08 Brush Welman Inc. (April 1976).
4. D.B. King, S.H. Gelles, and T. Nicholas, *Effect of Inclusions on the Mechanical Behavior of Beryllium.* AFML-TR-76-33 Brush Welman Inc. (April 1976).

5. G.J. London and W.G. Lidman, *Fabrication and Evaluation of Hot Isostatically pressed Beryllium.* AFML-TR-75-213 Kawecki Berylco Industries, Inc. (January 1976).

6. J.M. Finn, L.C. Koch, and D.L. Rich, *Design, Fabrication, Testing, and Evaluation of Damage Tolerant Beryllium Structures.* AFFDL-TR-68-108 McDonnell Company (August 1968).

7. G.R. Paul, *Fabrication Techniques of Beryllium Metal Removal.* S.A.E. 670803, Speedring Corp., Aeronautic and Space Engineering and Manufacturing Meeting, Los Angeles, California (October 2-6, 1967).

BIOGRAPHY

Leland (Lee) Norwood is a Member of the Technical Staff in the Producibility & Manufacturing Development Department at Rockwell International's Space Transportation and Systems Group. On the Shuttle program he has had the prime developmental and problem-solving responsibilities for all design producibility, fabrication, processing, and assembly involving beryllium applications, plasma-sprayed assemblies, window assemblies, and the thermal protection system thermal barriers and gap fillers. He has worked for Rockwell 46 years in mechanical and metallurgical engineering, advanced manufacturing development, and technical project management. Mr. Norwood studied mechanical and metallurgical engineering at the University of California, Los Angeles. He is the author of several reports and papers and holds one patent. He is a member of the Society for the Advancement of Material and Process Engineering and is a licensed Professional Engineer in Manufacturing Engineering, California, Certificate No. 3738.

LESSONS LEARNED FROM THE DEVELOPMENT AND
MANUFACTURE OF CERAMIC REUSABLE SURFACE INSULATION
MATERIALS FOR THE SPACE SHUTTLE ORBITERS

Ronald P. Banas, Donald R. Elgin, Edward R. Cordia,
Kenneth N. Nickel, Edward R. Gzowski, and Lawrence Aguilar
Lockheed Missiles & Space Company, Inc.
Sunnyvale, California

SUMMARY

Three ceramic, reusable surface insulation materials and two borosilicate glass coatings were used in the fabrication of tiles for the Space Shuttle orbiters. Approximately 77,000 tiles have been made from these materials for the first three orbiters, Columbia, Challenger and Discovery. Lessons learned in the development, scale-up to production and manufacturing phases of these materials will benefit future production of ceramic reusable surface insulation materials.

INTRODUCTION

The landing of Columbia after STS-5 on 11 November 1982 demonstrated the reality of a truly "reusable" thermal protection system. The concept of a non-ablating, rigid, reusable, ceramic insulation material was identified by a Lockheed patent disclosure in December 1960. It was recommended as a TPS for Lifting Reentry Vehicles by Lockheed in 1964 (ref. 1) and was pursued as a low level research and development effort during the early 1960's. A concentrated development effort was started in 1968 (refs. 2, 3, 4 and 5) to parallel the NASA Phase B studies that defined some early Space Shuttle configurations (ref. 6). Many lessons were learned during each phase of the evolution from laboratory development to an initial production facility in 1971 (refs. 7 and 8), and finally to the full production facility (refs. 9 and 10), which produced a shipset of tiles for the orbiter Columbia.

Lessons learned during the development and scale-up to production of three rigid, ceramic, Reusable Surface Insulation (RSI) materials and two borosilicate glass coatings will be discussed. However, the main emphasis will be on the significant lessons learned from the following manufacturing phases in the full production facility:

1. Processing of raw materials into tile blanks and coating slurries

2. Programming and machining of tiles using numerical controlled milling machines

3. Preparing and spraying tiles with the two coatings

4. Controlling material shrinkage during the high-temperature (2100-2275°F) coating glazing cycles

5. Measuring the tiles before coating and after coating glazing

6. Loading tiles into polyurethane array frames, shimming the tiles to the proper tile-to-tile gap width and machining the inner-mold-line of all tiles in an array

The RSI materials include LI-900 (Lockheed insulation at a density of 9 lb/ft^3), an all-silica material developed by Lockheed Missiles & Space Company Inc. (LMSC) in 1972. A predecessor, LI-1500, (a 15 lb/ft^3 density all-silica material) was developed by LMSC in 1962 (ref. 1). It was the lowest weight prime material for Lockheed's reusable lifting reentry vehicle studies (ref. 11) from 1962 until 1971 when LI-900 was developed (ref. 12). LI-2200, a 22 lb/ft^3 density all-silica material, was patented by NASA ARC (ref. 13) and scaled up to production by LMSC in 1977. FRCI-12 (Fibrous Refractory Composite Insulation at a density of 12 lb/ft^3) is a composite fiber RSI material. During design, development, test, and evaluation of Columbia, the need for improved thermal protection tiles was recognized. Stronger, less dense tiles more resistant to impact damage were desired. A ceramic tile material with these characteristics, in addition to the other required properties, was invented by the NASA ARC (ref. 14) and scaled up to production size billets by LMSC (ref. 15). This material, FRCI, composed of a blend of silica fibers and aluminum borosilicate fibers, is an outgrowth of LI-900 and LI-2200 technologies and basic research of high temperature materials.

The two borosilicate coatings are Class 1 and Class 2. The Class 1 (0036C) coating (ref. 16) is white and has a ratio of solar absorptance to total hemispherical emittance between 0.2 and 0.4 from -170°F to 135°F, and an emittance $\geq$ 0.7 at 1200°F. The Class 2 or Reaction Cured Glass (RCG) coating (ref. 17) is black, and has a total hemispherical emittance $\geq$ 0.8 at 2300°F and a ratio of solar absorptance to total hemispherical emittance between 0.7 and 1.1 from 170°F to 250°F. It is used on LI-900 tiles that experience surface temperatures from 1200°F to 2300°F and all LI-2200 and FRCI-12 tiles.

LI-900 was scaled up to production in 1975 and the first production billet for Columbia was fabricated in September 1976. LI-2200 was implemented as a pilot plant operation to produce about 100 tiles per orbiter in October 1977. After the final tile deliveries were made for Columbia and Challenger, about 3000 tiles per orbiter were made from LI-2200.

The pilot plant operation for FRCI-12 started in January 1979 under a contract from NASA ARC (ref. 15). Facility modification and the scale-up to production billet sizes started in October 1979. The first FRCI-12 production billet for Discovery (OV-103) was produced in October 1981. About 2700 FRCI-12 tiles are scheduled for installation on Discovery and Atlantis (OV-104). The processing parameters involved in the production of these materials are described in reference 18.

LIST OF SYMBOLS AND ABBREVIATIONS

APT   Applied Programmed Tool, a computer language used to drive numerically controlled milling machines

ARC   Ames Research Center

ATA   Array tile assembly, which consists of a polyurethane array frame loaded with tiles. Tile IML's are cut in the frame. The ATA is used as a shipping container and for tile installation on the orbiter.

Billet	A finished piece of LI-900, LI-2200 or FRCI-12
Breather Area	The uncoated area on the sides of the tile that starts at the coating terminator line and extends to the tile IML. The breather area allows air to vent out during ascent to preclude a loss of coating.
CADAM	Lockheed Computer Aided Design and Manufacturing system
CAD/CAM	Computer Aided Design/Computer Aided Manufacturing
CATIA	A Computer-Graphics Aided Three-Dimensional Interactive Application system developed by Dassault Aircraft in France
Class 1 Coating	White coating used for temperatures of 1200°F or less
Class 2 Coating	Black RCG coating used for temperatures between 1200° and 2300°F
Class 1 Tile	LI-900 covered on the OML and sides with a white borosilicate coating
Class 2 Tile	LI-900 covered on the OML and side with a black borosilicate coating (Class 2 coating)
Class 4 Tile	LI-2200 covered on the OML and sides with a black borosilicate coating (Class 2 coating)
Dry Density	The density of an LI-900, LI-2200 or FRCI-12 billet prior to exposure to the sintering cycle
FRCI	Fibrous Refractory Composite Insulation
FRCI-12	A rigid, composite fiber insulation made of 78% silica fibers, 22% Nextel fibers with 3% by weight of silicon carbide at an average density of 12.5 lb/ft^3
FRSI	Felt Reusable Surface Insulation
GHP	Guarded Hot Plate
IML	Inner Mold Line
IP	In-plane direction which is perpendicular to the through-the-thickness direction
LI-2200	A rigid, all-silica fibrous insulation with about 2% by weight of silicon carbide at an average density of 22 lb/ft^3
LI-900	A rigid, all-silica fibrous insulation with an average density of 8.75 lb/ft^3
LMSC	Lockheed Missiles & Space Co. Inc.
MD	Master Dimension

MDI	Master Dimensions Intersections
Mylars	Tile section cuts put on flexible heavy gauge plastic by Rockwell or LMSC and used by LMSC inspectors to measure the sides of complex tiles
NC	Numerical Control
Nested Tile	A tile that is individually measured for planform dimensions and has its IML cut while being held in a polyurethane nest
Nextel 312 ®	Aluminum borosilicate fiber; a product of Minnesota Mining and Manufacturing Co.
OML	Outer Mold Line; experiences aerodynamic heating during ascent and reentry
0036B	The original Class 1 coating; a dual layer coating consisting of porous optically adjusted subcoat and fused glass topcoat
0036C	The present Class 1 coating; a single layer system that meets the optical property requirements
0050	The original Class 2 coating; a fused silica subcoat and a topcoat of 7930 frit at 8% B_2O_3 with a silicon carbide emittance agent
PTX Lot	A blend of 20 Manville silica fiber lots
RCC	Reinforced Carbon Carbon
RCG	Reaction Cured Glass (the Class 2 coating)
RSI	Reusable Surface Insulation
STS	Space Transportation System
Terminator or Witness Line	The line that is put on the sides of most tiles to define the extent of the coating down the sides
TPS	Thermal Protection System
TTT	Through-the-thickness direction; also, the pressing direction during the casting operation

RSI LOCATIONS ON COLUMBIA

Over 30,800 RSI tiles were installed on Columbia by Rockwell. About 18,500 of the 23,400 tiles made by LMSC were HRSI, which is either LI-900 or LI-2200 with the Class 2 coating. The remaining 6000 tiles were LRSI, which is LI-900 with the Class 1 coating. The locations of the HRSI and LRSI tiles are shown in figure 1 along with the location of the RCC and FRSI. More details on the composition of these materials and the installation procedures used for all Orbiter TPS materials can be found in references 19 and 20.

LI-900 PROCESS DESCRIPTION

Materials

The principal component in LI-900 is all-amorphous silica fibers with an average diameter of 1.2 to 1.4 microns and lengths to 1/4 inch. During development, a major goal was to obtain a stable material that resists devitrification at elevated temperatures. This was accomplished in an extensive development program with the fiber supplier, Manville Corporation. The final product, Q-fiber, is amorphous silica with greater than 99.7 percent purity. These fibers retain their amorphous structure when exposed to a temperature environment of 2500°F for extended periods. The LI-900 system contains a colloidal silica binder that requires extensive treatment to obtain the purity required for high-temperature morphological stability.

Material Pretreatment

During the development of LI-900, certain pretreatment procedures were developed to improve uniformity and processability of the constituent materials. Maintaining uniform shrinkage characteristics was difficult early in the development of the process. This was overcome by heat-treating the fiber before processing it into billets. In addition, unfiberized glass called "shot", if not removed, causes high density, devitrified inclusions in the sintered material. To eliminate the "shot", the fiber is slurried with deionized water and passed through a hydro-cyclone cleaner (fig. 2). The cleaned fiber slurry is transferred into a centrifugal extractor to remove excess water and to form a fiber "cake", in preparation for final drying. Also, silica fiber lots received from Manville exhibit variable fiber characteristics that cause variations in billet densities. A blend of 20 Manville lots, called a PTX lot, was developed to reduce this variability (fig. 3).

LI-900 Fabrication

The LI-900 process flow is shown in figure 4. LI-900 billets are cast in two sizes, 15 x 15 x 6.5 inches and 10 x 20 x 7.3 inches. The operation is performed in an automated casting line. Preweighed amounts of fiber are loaded into twenty-six hoppers on a carousel that automatically positions and empties the hoppers sequentially. Originally, 4.9 lbs of fiber were used for each casting. Later development resulted in a change to 5.2 lbs of fiber along with a reduced water to fiber ratio (fig. 5). These changes, plus others to be described later, resulted in improved density distribution within the billets. The fiber and a pre-determined quantity of water are combined in a tank containing a low shear mixer that uniformly disperses the fiber with minimum chopping. At the conclusion of the timed mix cycle, the slurry is automatically transferred into a casting mold positioned directly below the mixing tank.

Entrapped air bubbles are removed from the slurry prior to compressing the billet to its final cast size. This is accomplished by a combination of vibration and stirring. Care must be exercised during this operation to maintain a homogeneous dispersion of the fiber. Water is removed and the casting is compressed to a specified height in the volume adjustment operation. Concurrently, a vacuum is applied to the bottom of the mold to remove a specified quantity of water. At this stage, the standard billet contains 5.2 lbs fiber and approximately 24 lbs of water. The next step in the casting operation is the dispersion of a colloidal silica binder in the compressed casting at the weight adjustment station. The binder solution components are automatically mixed and dispensed through metering pumps. The solution is dumped on top of the compressed billet and a vacuum is applied to the bottom of the casting mold. The residual water in the billet is displaced by the binder that is dispersed throughout the casting to a specified solids concentration. Upon removal from the mold, the wet billet is weighed to provide a check that all steps of the casting operation were performed correctly. The current method, described above, is an improvement over the original method which excluded the void reduction and vacuum water withdrawal (fig. 6).

Since maintaining a uniform distribution of the colloidal silica binder in the casting is important to maintain uniform physical properties, a gelling agent is used to set the binder and prevent it from migrating during drying. The billets are dried using either a conventional oven or a microwave dryer. They are weighed after drying to assure that the specified amount of water is removed prior to sintering. Originally, the castings received a first and second sintering with an additional binder addition between sinterings (fig. 5). Later development resulted in a single sintering combined with the change from 4.9 to 5.2 lbs of fiber. The result is an improvement in billet density distribution and an increase in yield.

An additional improvement in average billet density was obtained with the implementation of a fiber compact shrinkage test. A correlation between billet density and sintering schedule was developed for each PTX lot (fig. 7). This allows adjustment of the sintering schedule to accommodate the PTX lot shrinkage characteristics which influence the billet density.

Originally, the dried castings were sintered in specially designed 3-zone tunnel kilns at a peak temperature of 2350°F. These kilns were used from 1975 to 1982. Early in 1982, the sintering operations were transferred to elevator kilns. Six side heating is utilized in these kilns compared to five side heating in the previous kilns (fig. 6). This improves the strength distribution within the billets. The sintering schedule is adjusted to produce billets with an average density of 8.8 lb/ft^3 by adjusting the sintering time to accommodate the PTX lot shrinkage variations.

LI-2200 PROCESS DESCRIPTION

Materials

LI-2200 is composed of amorphous silica (Q-fiber) and a small amount of silicon carbide powder which provides additional thermal protection if the material is exposed to excessive temperatures due to coating loss at the tile outer surface. With LI-2200, the fiber heat-treatment is omitted, and only the hydro-cyclone cleaning is performed. Until January 1981, the fiber cleaning was performed by an air bubbling procedure which was less efficient and less reproducible than the present procedure.

LI-2200 Fabrication

The LI-2200 process flow is shown in figure 8. The billets are cast in a specially designed, manually operated casting tower in a 14.4 x 14.4 x 8 inch size. The mixing process differs from LI-900 in that the preweighed fibers are combined with water, SiC, and ammonium hydroxide into a V-blender equipped with an intensifier bar. Since LI-2200 requires a significantly higher casting density than LI-900, the slurry requires some chopping action to obtain the necessary fiber packing. After the blended slurry is transferred into the casting tower and sealed, void elimination is accomplished by applying a high vacuum to the slurry prior to billet formation. The billet is formed by removing part of the water by gravity drain, compressing the slurry to a final thickness, and then extracting additional water with vacuum.

The billets are dried in a batch oven at 450°F for 16 hours. The dry density of the LI-2200 is approximately 13 lb/ft^3. Originally, this was the final operation before sintering. However, the billets sometimes exhibited cracks after sintering. An additional drying at 1000°F for 12 hours was developed and implemented in September 1981 to eliminate this problem.

The LI-2200 is sintered in elevator kilns. The sintering schedule is similar to that used for LI-900, except that the peak temperature is 2420°F. The soak time at peak temperature is adjusted to maintain final densities within a 22 ±2 lb/ft^3 density range. Originally, the soak times were based on fiber chemistry. A more accurate method, based on fiber compact shrinkage, was developed and implemented in June 1981.

FRCI-12 PROCESS DESCRIPTION

Materials

FRCI-12 is a composite fiber material containing amorphous silica (Q-fiber ®) and aluminum borosilicate fibers (Nextel 312 ®, a product of Minnesota Mining and Manufacturing Company) in a fused fiber matrix. Silicon carbide powder is added for additional thermal protection as it is in LI-2200. The bulk silica and Nextel fibers are heat treated at 2200°F and 2000°F respectively to stabilize and standardize fiber properties. Hydro-cyclone cleaning is performed on the silica fibers to remove particulate contaminants, followed by drying.

Fabrication

The FRCI-12 process flow is shown in figure 9. Silica and Nextel fibers are intermixed and cast into billets using a multi-stage, wet-slurry blending process and automated casting equipment. Castings are dried using a combination of microwave and convection-air ovens to achieve optimum drying rates. Dry castings are sintered in elevator kilns using microprocessor controllers to achieve uniform, repeatable heating to the optimum sintering temperature (approximately 2400°F). The optimum sintering temperature varies as a function of FRCI composition and desired final density.

FRCI can be fabricated with a range of compositions and densities to allow tailoring the material to a specific application. An FRCI formulation with a density of 12 pounds per cubic foot and a silica to Nextel fiber ratio of 78/22, identified as FRCI-12, was developed to replace LI-2200. Other FRCI materials with densities of 8 lb/ft^3 and 20 lb/ft^3 have been produced. FRCI-12 tiles were substituted for approximately 2764 LI-2200 tiles on the third orbiter, Discovery. Substituting FRCI-12 for LI-2200 saves approximately 870 pounds of weight per orbiter due to the lower density of FRCI-12. Also the tensile strength design value is increased by 50 percent, and the susceptibility to coating impact damage is reduced by eliminating residual tensile strain in the coating due to a better match in coefficient of thermal expansion between FRCI and tile coating materials.

Development of full-scale manufacturing processes for FRCI-12 required considerable effort by many individuals within NASA, Rockwell International, and LMSC from October 1979 through October 1981. Several important lessons were learned during this development about the inter-relationships between processing and fundamental material properties. The development effort was complicated by the requirement to produce a tile material to meet all the existing requirements of the baseline material, while also providing improvements of significant importance to warrant replacement of proven materials.

The first significant FRCI problem was encountered during initial scale-up work on the NAS2-10134 contract. Nextel fibers did not readily disperse when blended with silica fibers in the full-scale mixing equipment. Clumps of undispersed Nextel fibers, which varied from 1/8 to 1/2 inches in length and were present in the sintered FRCI-12 material, caused unacceptable coating discontinuities on finished tiles. The original laboratory method called for wetting the Nextel fibers prior to introduction into a small lab-scale V-blender containing silica fibers. This lab-scale equipment and procedure produced an acceptable mixture of the two fibers in lab-size castings. However, when the same procedure was used in full-scale production equipment, the Nextel fibers did not disperse and Nextel clumps occurred in the finished material. An interim dispersion method, only marginally acceptable, was devised for the pilot plant operation conducted under NAS2-10134. Nextel fiber was preblended in water in a lab-size V-blender to break up Nextel clumps, mixed with an equal amount of silica fiber in the same blender to prevent reaggregation of the Nextel into clumps, and then blended with the remaining silica fiber in the full-scale blender to produce a slurry with suitable characteristics for casting. This interim dispersion method reduced the number and size of clumps in the finished material, but the method was not considered suitable for full-scale production due to the need for considerable coating touch-up. A high-speed, high-shear blender was substituted for the small V-blender for preblending operations in the final production process. As shown in figure 10, use of the high-shear mixer for preblending Nextel fibers totally eliminated Nextel clumps in finished tiles (ref. 21).

The second significant problem was encountered during characterization testing of the pilot-plant material. The apparent[1] thermal conductivity of the pilot-plant FRCI-12 was higher than the LI-900 base line value. Rockwell's criterion for substitution of any material for LI-900 or LI-2200 was that the thermal response must be equal to or lower than that of LI-900. A guarded hot plate (GHP) apparatus is normally used to characterize the thermal conductivity of a material. Measurements can be obtained over a wide range of temperatures and pressures to establish design values. However, the $\pm18\%$ uncertainty band associated with GHP data makes comparative measurements on different specimens, with minor variations in thermal conductivity, uncertain. Comparative measurements are more easily accommodated with the instrumented tile method (ref. 23) shown in figure 11. Tiles fabricated from different materials can be tested side-by-side in a radiant heating environment at reduced pressures. Either steady-state or transient heating conditions can be simulated. This method yields more reliable comparative results than the GHP method which is limited to testing one material at a time. Apparent thermal conductivity values for laboratory FRCI-12 were much lower than the pilot-plant FRCI-12, but still higher than LI-900, indicating that some key parameter(s) must have been inadvertently varied during scale-up. An investigation of the effects of various compositional and processing variables on FRCI-12 properties showed that apparent thermal conductivity can be affected by several factors, the most important being density change during the billet sintering cycle (ref. 24). Pilot-plant FRCI-12 experienced a density change of 5.5 lb/ft^3 during sintering, whereas laboratory material experienced a change of 4 lb/ft^3. Full-scale production FRCI-12, which has acceptable apparent thermal conductivity (ref. 25), experiences only a 2 lb/ft^3 density change. Reducing the density change during sintering was accomplished by use of the high dry density concept, producing castings with increased fiber content (i.e. more fibers per unit volume) and sintering at a lower temperature for a shorter time (fig. 12). Reducing the time/ temperature profile during sintering caused a reduction in average tensile strength of the material compared with pilot-plant material. However, use of six-sided heating in the kiln in place of the five-sided heating resulted in a more uniform strength distribution within the billets and provided design tensile strength values nearly equal to the pilot plant values by lowering the deviation.

Another significant development problem was encountered when attempts to achieve FRCI-8 (8 lb/ft^3) thermal conductivity equivalent to LI-900 were unsuccessful. Minimizing the change in density during sintering was not sufficient to produce FRCI-8 with acceptable thermal conductivity. A combination with other, less significant, thermal conductivity "drivers" was necessary. Experiments showed that reducing the Nextel fiber concentration in the material formulation and reducing the size of the silicon carbide particles in the material, provided the additional reduction in thermal conductivity that was required (ref. 26). Pilot-plant FRCI-8 experienced a change in density during sintering of 2.5 lb/ft^3, has a silica to Nextel fiber ratio of 78/22, and contained 320 mesh silicon carbide particles. Full-scale FRCI-8, which has acceptable

---

1 For porous materials, the term apparent thermal conductivity is used to denote heat transfer within the material by solid conduction, gas conduction and radiation (ref. 22).

thermal conductivity, experiences only a 1.5 lb/ft^3 change in density, has a silica to Nextel fiber ratio of 85/15, and contains 600 mesh silicon carbide particles. Reducing the change in density during sintering was accomplished by producing castings with higher dry density and sintering at a lower temperature for a shorter time (fig. 13). Reducing the time/temperature profile during sintering and reducing the concentration of Nextel fibers in the formulation resulted in lower average strength for the full-scale material compared with the pilot-plant FRCI-8. However, use of six side heating in the kiln instead of five side heating, and increasing the heat-up rate to the sintering temperature (to minimize shrinkage during heat-up and maximize time above the critical fiber bonding temperature of 2350°F) resulted in a more uniform strength distribution within the billets and provided design allowable strength values nearly equal to pilot plant values.

TILE PROCESS FLOW

The sequence of operations performed after the insulation material is cut into cubes is shown in figure 14. After the tiles are machined, they are heat cleaned to remove organic contaminants, masked to allow an uncoated breather space area along the lower perimeter adjacent to the IML, sprayed with Class 1 or Class 2 coating and sintered at 2100 to 2250°F in an Ipsen tunnel-hearth roller kiln. After vacuum waterproofing with a methyl trimethoxy silane (Dow Corning DC 6070), the tile identification number is painted on the OML and the tiles are checked dimensionally as required prior to the IML cuts.

The tile IML cut is performed either individually, which is called a "nested" tile, or on a group of tiles simultaneously in an array frame, which is known as an ATA.

The dimensions of a nested tile are checked as required prior to the IML cut. The dimensions of the ATA tiles are checked by their ability to fit into a premeasured polyurethane array frame with the required tile-to-tile gaps, which are generally 0.045 ±0.016 inch on the lower wings and fuselage and 0.055 ±0.016 inch on the upper wings, fuselage and vertical fin.

ENGINEERING

Engineering Data Flow

Engineering data, which is used to define the tile and array frame geometries, is received from Rockwell in the form of engineering assembly drawings, tile bounding plane data and inner/outer mold line data (fig. 15). Only 115 of the 23,400 tiles that Lockheed made for OV-102 are defined by conventional engineering drawings. The mold line data can be represented by points (X, Y, Z coordinates) and corresponding normal vectors (MDI data), and recorded on a magnetic tape and/or contained as surface definitions described in the Master Dimensions Specification Book, Document No. MD-V70, Rockwell International. The MDI data are transformed from an orbiter coordinate system to a local tile/array coordinate system that is compatible with the APT language. These transformed data are stored in a geometry file that is accessed by the NC programmer for use in preparing the part program to machine tiles.

A tile machining drawing is used to determine the proper tile shrinkage compensation factor (see Tile Measurement and Shimming Methods section) to include in the tile part program.

The geometry file, which contains the tile boundary planes and the OML and IML surfaces, is also used to write the NC part program to machine the array frame. LMSC fabricated 739 array frames for Columbia. Product Assurance Inspection Standards are also prepared for use in inspection of the tiles on the Cordax measuring machines.

Master Dimension Refinements

Almost 18,500 of the 23,400 tiles LMSC made for OV-102 were MDI tiles, which are defined in a grid of X, Y, Z coordinates and corresponding normal vectors. The remaining tiles are the more complex MD tiles, which have their geometries defined in the Master Dimensions Specifications book. Substantial refinements have occurred in the procedures used to define the 4900 MD tiles. Originally, hand calculators or personal computers were used to calculate points to approximate complex surfaces, tile and array corner points, Product Assurance Inspection Standards, and points to check the accuracy of inspection aids (mylars) furnished by Rockwell (table I). Surfaces which could not be analytically defined were approximated by calculating points and passing a curve through these points.

Software, in the form of APT and FORTRAN computer programs, is now being used to calculate tile and array corner points and to provide accurate blank sizes. Additional software and a CAD/CAM system are both used to develop and check the accuracy of complex surfaces. This same CAD/CAM system is used to provide mylars to check hard-to-inspect tiles, reducing the time required for tile inspection. The accuracy of certain Rockwell furnished mylars is checked at LMSC using the CAD/CAM system if a discrepancy is noted during the inspection process.

Mylars were seldom used for Columbia tiles. However, for Challenger and Discovery, mylars were used extensively for hard-to-measure tiles. For example, 123 complex hinge cover tiles, which contain conical surfaces, ruled surfaces and through holes, were recently delivered for Discovery ahead of schedule. This success was due to a joint LMSC-RI effort to make 28 mylars that were used to inspect these tiles.

NC Programming Refinements

Initially, an attempt was made to automate tile programming by using a cut package to program each family of tiles (table II). Each minor difference in tile geometry required a different cut package which was inefficient and often difficult to use. Limited knowledge of the unusual and complex surfaces involved made the programming task very difficult and time consuming. Programmers experienced many failures before being able to visualize a tile, working from just the master dimensions data. It required 90 NC programmers working for ten months, using more than 160 hours per week of Univac 1108 computer time. An interactive graphics system was not available to program tiles.

Tiles with planar sides were programmed as planar surfaces. This resulted in corner shrinkage when the tile coating was glazed (see Tile Shrinkage section). One cutting tool was used to machine most tiles. Since no coating terminator line was machined on complex tiles, problems were encountered when these tiles were masked for spraying and a high rejection occurred for tiles with insufficient breather area (fig. 16). The use of different fixtures to locate different size blanks on the three Danley Corp. NC machines caused a relatively high percentage of tiles to be scrapped due to operator error in locating the blank.

Programming and tile machining are now more efficient due to knowledge and experience gained over the life of the program. About 6000 tool tries were required to develop the proper part programs for the tiles on Columbia. About 1400 tool tries were required in connection with design changes on Challenger and about 600 tool tries as a result of design changes were required on Discovery. Planar sides are now programmed as cylinders with 900-inch radii to reduce tile corner shrinkage. About 30 electroplated diamond tools ranging in diameter from 1/8 to 2 inches were designed during the first year of production and are now used to reduce machining time. A coating terminator line is now added to most complex tiles to facilitate coating spraying and reduce the number of tiles scrapped or reworked due to incorrect breather area. The same fixture is also used on the NC mills to locate all blanks, regardless of size. This change has greatly decreased the number of tiles scrapped because of operator error in locating the blank on the bed of the NC machine.

Interactive Graphics

The recent use of an interactive graphics system (CATIA) to program the redesign of specific complex tiles has demonstrated that this method of programming reduces the number of tool tries required before an acceptable part can be made. The NC programmer has the ability to replay the cutter motion on the graphics system terminal and correct any errors observed prior to machining the first tool try.

The expanded use of an interactive graphics system for the redesign of the more complex HRSI tiles would markedly reduce both the cost and time required to manufacture these tiles (fig. 17). An example of how this system would work follows. A three dimensional engineering model of a given tile is constructed on the system. The tile model can be rotated on the terminal scope, showing all facets and all surface intersections. The model is then accessed, the cutter motion is added, the cutter motion is replayed to check for and correct errors, and then a tape is produced and sent to the machine shop for a tool try. Any errors encountered during the tool try can be easily corrected using the same engineering model. However, as the NC programmer becomes more proficient with the 3-D model, this sequence should minimize the number of tool tries. In addition, Product Assurance can access the same engineering model and extract the attributes necessary to inspect the tile.

BOROSILICATE COATINGS

Class 1 Coating

Development of the Class 1 (white) coating was a significant challenge because of the stringent optical property and weight requirements. The optical property requirements are a ratio of solar absorptance to total hemispherical emittance between 0.2 and 0.4, to achieve low temperatures while on orbit, and an emittance of 0.7 at 1200°F to allow maximum re-radiation of the convective heating energy during reentry (fig. 18). An intensive development program was successful in producing a dual layer coating that was started in production in October 1977. This coating (0036B) consisted of a fused, water-impervious topcoat of clear glass plus zinc oxide, over a porous subcoat that contained aluminum oxide for high reflectance and silicon carbide for high emittance. The subcoat required drying at 1300°F prior to spraying the topcoat. After both layers were applied, glazing at 2100°F was required to produce a water-impervious, dual layer coating.

In mid 1978, effort was directed toward combining the dual layers while retaining both the optical properties and the water imperviousness. A single layer coating (0036C) was successfully developed, qualified and implemented into production in early 1978. During this period, the maximum coating weight requirement was increased from 0.09 to 0.12 lb/ft^2 to alleviate a coating cracking problem which was unavoidable with a 0.008 inch thick coating. This coating was successfully applied to about 5700 tiles for orbiter 102. For Challenger, OV-099, the maximum coating weight requirement was increased to 0.17 lb/ft^2. While in production for OV-102, a major water imperviousness problem affected the Class 1 coating. A three month investigation revealed that the cause was large frit particle size that precluded complete fusion (ref. 27). A complete particle size distribution requirement was determined and imposed on the frit supplier, Corning Glass Works. Particle size controls were also instituted for the coating slurry (fig. 19) to assure complete fusion during the 2100°F coating glazing cycle.

Class 2 Coating

The Class 2 (Gray) coating (0050), developed in 1974 (fig. 20), was a dual layer system that contained Corning 7930 frit with a boria content of 8% and a silicon carbide emittance agent. Because of a high coating residual tensile strain (values of 200-300 microinches per inch), crack propagation was not inhibited. This coating was replaced in June 1976 with a NASA ARC-patented Reaction Cured Glass (RCG) coating. The RCG coating is a single layer system that meets all the optical property requirements and also has lower coating residual tensile strains. It was implemented into production in May 1976 and was used on 195 tiles that were installed on the lower mid-fuselage of the Enterprise, which was used for all the subsonic aerodynamic tests at the NASA Dryden Flight Research Center. Subsequently (ref. 28), the RCG coating was shown to be susceptible to both coating impact damage and crack propagation. However, this susceptibility is probably common to any thin glass coating; RCG is a single layer coating system that was easy to scale up to a production operation and which proved to be repairable when damage occurred.

The Class 2 frit used in the RCG coating also encountered a particle size anomaly in January 1976 when the coating process was transferred to LMSC from NASA ARC (ref. 29). Examples of coating anomalies that are caused by too many fines (particle size less than 1 micron) are shown on the left side of figure 19. Within 3 months after the implementation of both the Class 1 and Class 2 coatings into the production facility, both frits and slurries were controlled by full particle size distribution requirements (ref. 30).

Universal Patch

Rigidized fibrous insulation is subject to casting voids and to scratches and gouges during handling. The initial method for filling these voids was to fill them with cured silica slip for Class 1 coated tiles and with RCG coating for Class 2 coated tiles. This process required multiple fill and drying cycles. Also, there was concern that the dense fills could vibrate loose, and further enlarge the voids.

A universal patch material was developed that has a density approximately equal to that of the tile, is applicable to both silica and FRCI tiles, is compatible with the coating glazing cycle, is capable of repairing tile edges and corners, and is easily and rapidly applied. Existing, approved Shuttle materials which are used to compound the patch material are silica fibers, colloidal silica, acrylate solution, deionized water, and a combination of fuchsin and methyline blue dyes. The slurry is simply placed into a void at twice the void volume, and flattened with a teflon spatula. After patching, the tile is dried at 1200°F for 8 minutes, and is then ready for coating. Full patch cure occurs during coating glazing. Excellent bonding of patch to tile has been demonstrated, and no crystallization occurs from exposure to a temperature of 2300°F for 15 hours. The scrap rate for damaged tiles was significantly reduced after this procedure was introduced into production.

Tile Coating Application

Class 1 and Class 2 borosilicate glass coatings are applied to tile blanks by spray application of a slurry. Tile blanks are set up on holding fixtures, masked to provide a breather space near the IML surface, and patched as necessary to cover surface deformities. Class 1 tiles are seal-coated with a suspension of colloidal silica in water and dried prior to application of the coating. Class 2 tiles are wetted with alcohol prior to application of the coating.

The amount of slurry applied to each tile is controlled by maintaining slurry viscosity, line pressure, and the number of coats within predetermined limits. The coating weight is determined "wet" and a conversion factor is applied to calculate "dry" weight. Coating weights are between 0.07 and 0.17 lb/ft^2 for Class 1 tiles and between 0.09 and 0.17 lb/ft^2 for Class 2 tiles. The corresponding coating thicknesses are 6 to 15 mils for Class 1 tiles and 8 to 15 mils for Class 2 tiles.

Several significant problems with the coating process were encountered during production of tiles for Columbia. One problem involved robot sprayers, which were initially used in 1977 to coat the less complex tiles. The first 3000 to 4000 tiles, primarily Class 2, that were coated using the robots had excessively high reject and/or rework rates for coating deficiencies such as runs, non-uniformity and excess or insufficient coating thickness. Extensive experimentation with adjustments and programming of the robots indicated that the following problems could not be resolved with the existent capabilities of these first generation robots:

1. They were unable to accommodate minor variations in viscosity typically encountered with production batches of slurry.

2. There was no capability to change the speed of the robot from that used during the programming (teaching) phase.

3. Robots used twice as much slurry as manual spraying (i.e. only 45 tiles per 5 gallon batch were sprayed compared to as many as 80-100 tiles per batch for manual spraying).

4. The cassette tapes used to control the robots were not interchangeable between robots so each robot had to be taught (programmed) individually.

5. Dirt on the tape heads caused unplanned and uncontrollable motion in the robots.

6. There was no feedback loop in the system during the production spraying phase that could change the speed or the rate at which the slurry was being sprayed.

It was found that experienced coating technicians could accommodate the variations in tile geometry and slurry viscosity and provide a yield in excess of 95% for this operation (fig. 21). Use of robots was discontinued for spraying production tiles in March 1979. Current, more sophisticated robots, with advanced technology such as control by floppy disk or computer, and active feedback loops, could probably handle the mechanical problems. However, the ability to distinguish subtle changes in slurry viscosity and apply in-process corrections still appears to be handled best by skilled operators. A qualified sprayer can adjust the spraying speed to overcome any subtle changes in viscosity and can also touch up the tile as required at any time in the spraying sequence.

Another significant problem was high rejection rate for coating weight discrepancies. Coating weights were initially determined after glazing. Excess or insufficient coating caused the tile to be scrapped. A method of determining the coating weight while still "wet" was developed. The "wet weight" is determined by the operator and corrections are made if necessary before the coating dries. Underweight tiles receive an extra coat of slurry and overweight tiles are stripped and recoated (fig. 22). Another problem was changes in slurry viscosity with time. Slurry viscosity degraded with time to the point where it was too "thin" to be sprayed without running and sagging. Investigation showed that trace amounts of iron contamination were introduced by processing equipment at the frit manufacturer's production facility. The iron oxidized with time in the made-up slurry at LMSC, destabilizing the particle suspension. A heat treating procedure at 1000°F was used initially to oxidize the iron before making the coating slurries. Later, the source of the iron contamination at the manufacturer was identified and eliminated.

Another significant problem was wide variation in tile shrinkage rate and degree of coating fusion. Investigation showed that a narrow and repeatable temperature range is required to provide the desired dimensional tolerances and uniform degree of fusion. The degree of fusion not only affects appearance, but also water imperviousness. The original glazing kilns (stage tunnel kilns) were not capable of holding the desired temperature tolerances. Tunnel hearth roller kilns (Ipsen Inc.) were obtained that maintain temperatures within ±10°F between 2100 and 2300°F. Tile glazing problems were virtually eliminated by using these kilns.

Tile Waterproofing

Initially, the tiles were rendered hydrophobic (waterproof) by immersion in a hexamethyl - disilazane silicone/freon solution. Tile weight gain was about 1.0 percent. Several hundred grams of freon solvent evaporated from each tile during thermal exposure. The process provided good water imperviousness, but left a dark, carbonaceous residue when heated to between 800° and 1000°F, resulting in an increase in the ratio of solar absorptance to hemispherical emittance to above the 0.4 specification limit.

The present method involves release of a trimethoxysilane vapor into a vacuum chamber containing the tiles. The tile weight gain is only 0.1 percent. Also, the material sublimes in a char-free condition, and no change in optical properties is encountered.

Tile Shrinkage

During the development of LI-900, dimensional control had not been identified as a problem since little was known about tile gap heating, and the tile dimensional tolerances were not defined. Tile shrinkage and warpage were not fully understood and the significant factors that influence tile behavior during the glazing cycle were not known.

In 1976 a series of test programs led to the realization that tile shrinkage could be correlated to a single factor, tile thickness, if all other parameters were held constant. This knowledge was aided by the utilization of precision Bendix Cordax machines, which provide accurate, repeatable measurement data for each tile. This enabled Lockheed engineers to develop a clear and complete picture of tile dimensional changes during glazing.

The initial results showed that tile length or width changes were related to the glazed tile thickness. A best-fit, least-squares logarithmic equation was developed using the available data. Reference 31 describes the details of this activity.

A plot of the basic offset equation, which reflects the use of the Richmond III glass melt, is shown in figure 23. Notice that the curve crosses the dashed line of zero shrinkage at a sintered tile thickness of 2.1 inch. This means that "thin" tiles experience a net shrinkage, while thick tiles "grow" due to addition of the coating on the sides.

The basic offset equation was used for most of the 15,000 NC tile part programs that Lockheed wrote for OV-102. These programs are relatively expensive software and are not easily changed. After a new glass melt, Waterville 1, was put into use, it was discovered that tiles made from the new glass melt fibers did not shrink the same as tiles from the original Richmond III glass melt. Therefore, another test program was conducted and a new offset equation was developed.

A plot of this new equation is shown in figure 23. Waterville 1 tiles shrink more than Richmond III tiles. The difference is significant when compared to the allowable side tolerance of ±.008 inch. Since it was not cost-effective to revise the NC part programs to correct for the increased shrinkage of the new melt, it was decided to continue using the existing NC software and to apply an offset correction at the time the tile is machined. Figure 23 shows the offset for a given glazed tile thickness. A table of tile machining offsets was developed and implemented by entering a letter code on the IBM travel card that accompanies each tile. New software was written which allowed the NC machine to "read" the letter code and apply the corresponding offset to each tile side during machining.

As new glass melts were introduced into the manufacturing process, test programs were conducted to develop the appropriate offset tables (ref. 31). To date, seven melts have been used for all orbiters and an eighth melt is being processed. Approximately 110,000 tiles have been made from these melts to date.

TPS tiles were originally machined with planar vertical sides. Subsequent shrinkage investigations revealed that special offsets were necessary to control shrinkage during coating glazing (fig. 24). These special offsets are classified into three categories: planar, which is the type of shrinkage discussed above, side-slope and radius-type compensation. The planar type shrinkage, which is the largest of the three, accommodates planform shrinkage during the glazing cycle. The side-slope and radius compensations are smaller in magnitude and accommodate distortion shrinkages. The side-slope distortion occurs because the OML edge shrinks more than the base of the tile. Radius compensation is necessary because the corners shrink more than the middle side of the tile. All adjustments are made in an equal and opposite sense and constitute typical adjustments made to NC machine part programs.

The solution to the side slope distortion problem was simple and economical. Since the original part programs used the coating terminator line as the drive path, a conically shaped tool having the same diameter at the tip as the cylindrical tool was designed (fig. 26). The cone angle was designed to give optimum offset to a majority of tiles that were manufactured by LMSC. With the same diameter at the tip as the cylindrical tool, the same part programs could be used with no changes.

The radius type compensation is made by simply machining a cylinder through a point at the center of the tile side. This results in more material left at the corners to accommodate the "pillow" type distortion.

During the early stages of Columbia tile delivery, simple (square-flat) tiles were manufactured. When more complicated tiles were fabricated, dimensional problems occurred. Tiles whose sides were not parallel or normal to each other (i.e., tiles with a wrap-around OML) had a high dimensional rejection rate. An investigation revealed that the shrinkage normal to the in-plane direction, which is defined as the through-the-thickness direction, is about three times larger than the in-plane shrinkage (fig. 27). The in-plane direction usually denotes a plane that is parallel to the orbiter surface. The shrinkage and distortion of these surfaces follow a complex relationship. Since the numerically controlled machines have limitations in application of automatic offsets to only the in-plane direction, which usually lies parallel to the machine bed, no letter offset method exists to adjust the part programs. As a result, a fixed offset is used in the NC part program for through-the-thickness shrinkage corrections. For example, all elevon cove tiles receive a fixed through-the-thickness offset for a specific silica glass melt. Wraparound and step tiles are treated in a similar manner (fig. 27).

Tile Measurement and Shimming Methods

Most of the OV-102 tiles had to be measured and verified for planform dimensional conformance prior to loading into array frames to machine the IML (fig. 14). Consequently, an automated system to measure tiles was implemented. Two Cordax measuring devices were programmed to automatically summon the inspection standards from a host computer, locate the tile on the machine bed and automatically determine the acceptability of the tile by using a series of 6 to 12 predetermined touch points on the tile sides, and 5 touch points on the OML surface.

The average time to measure a tile using a Cordax machine is 5 to 15 minutes. Another device, the "maxi-measure" (fig. 28) was developed to reduce the load on the Cordax measurement machines. This device consists of two parallel plates that measure the maximum dimension of tiles with parallel sides. The average measurement time for the "maxi-measure" device is less than one minute.

After about 70% of the tiles were fabricated for OV-102 the "load-and-go" concept was implemented to reduce the time required to dimensionally inspect tiles and to increase the rate of ATA deliveries to Rockwell. As shown in figure 29, the concept consists of an initial measurement of the array frames with aluminum templates or by probe on a large bed NC mill. The tiles are then loaded into the frame and shimmed to the proper gaps. If the proper tile-to-tile gaps are obtained, the IML's of all tiles in the array are cut as a group on one of the two large bed NC mills. The ATA's are then shipped to Rockwell. The advantages of the "load-and-go" concept are:

1. The number of tiles that must be checked for planform dimensions is greatly reduced.

2. Tile planform dimensional outages greater than ±.015 inch per side are allowed but the proper tile-to-tile gaps are maintained, and the overall array dimensions are to print.

3. ATA's that did not shim to the minimum tile-to-tile gap were reworked by refiring an entire row of oversized tiles to reduce their planform dimensions (Fig. 30).

4. If the ATA cannot be shimmed to the maximum tile-to-tile gap selected, tiles are remade to allow the ATA to shim properly.

For OV-102, about 3600 tiles were shipped to Rockwell in 170 ATA's using the "load-and-go" concept. For OV-099, which was the first shipset to use the "load-and-go" concept for all AFA's, 3,200 tiles were shipped as nested tiles and 20,500 tiles were loaded into 745 ATA's under the "load-and-go" concept. For Orbiters 103 and 104, which have about 18,200 LMSC tiles, about 2,100 tiles are nested and 15,800 tiles will be loaded into 535 arrays under the "load-and-go" concept.

Material Physical Properties

LMSC has had responsibility for material characterization tests of all the ceramic RSI materials. Rockwell has had responsibility for performing the systems tests on all RSI materials including RCC and FRSI. Figure 31 shows typical average room temperature physical properties developed by LMSC for

LI-900, LI-2200 and FRCI-12. More detailed data along with values at both elevated and cryogenic temperatures can be obtained in reference 19.

CONCLUDING REMARKS

This paper has presented a multitude of lessons learned in the development and scale-up to production of three ceramic RSI materials and two borosilicate coatings that were used in the manufacture of tiles for the first three orbiters: Columbia (late 1976 to early 1979), Challenger (April 1979 to March 1982), and Discovery (March 1982 to present). These improved methods, which are summarized in Table III, are presently being used in the fabrication of tiles for Atlantis, the fourth orbiter.

The effectiveness of the lessons learned is revealed in the overall tile yields: 48% for about 23,400 tiles for Columbia, 81% for about 23,800 tiles for Challenger and 88% for about 18,200 tiles for Discovery. With the deletion of certain planform measurements of nested tiles on 1 February 1983 as a result of an expanded process control program, the overall yield on Atlantis tiles is expected to be about 90%. While some of the increase in yield can be attributed to modified requirements, the majority of the yield increase is due to the improved methods discussed herein, primarily the addition of coating terminator lines on most tiles, the addition of homing devices on the NC mills, the reliance on real time process control for coating weight, and the "load-and-go" concept.

Experience since the start of production in October 1976 has shown that ceramic fiber reusable surface insulations still retain some "art" in their fabrication processes as opposed to all "science". Consequently, making a consistent, repeatable product requires good process control and all changes to the process must be thoroughly evaluated prior to implementation and tightly controlled after implementation.

Another lesson that has been illustrated through the development of LI-900 (ref. 12), LI-2200 (ref. 13) and FRCI-12 (ref. 15) is that the RSI materials can be "tailored" to the application as with fiber-reinforced composites. This "tailoring" is also evidenced in the recent advanced studies of FRCI using ratios of Nextel to silica of up to 80/20 (ref. 32). Hence, these families of RSI materials offer the designer a very flexible design concept.

Finally, if LMSC were to introduce a new, man-rated ceramic RSI material into production for an advanced Shuttle or Orbital Transfer Vehicle, the minimum changes that would be introduced are:

o Vacuum degassing of casting slurries

o Addition of silicon carbide particles to the billets for emittance retention of the RSI in the event of coating loss during entry

o Use of an interactive graphics system like CATIA or CADAM for the design of tile geometries and to provide an automated method to write NC part programs

o Implementation of real-time process control in critical manufacturing areas

o Consideration of different coating and tile concepts if rewaterproofing is required after every flight (i.e., tiles with larger planform dimensions and fewer if any material shrinkage problems in the absence of a coating glazing cycle)

REFERENCES

1. Thermal Protection Concepts for Lifting Entry Vehicles. Lockheed Missiles & Space Co. Report 6-62-64-5, May 5, 1964.

2. Hammitt, R. L.: Lightweight Insulation, LI-15, Test Summary. Lockheed Missiles and Space Co., LMSC-685434, Code 9999, Apr. 26, 1968.

3. Rusert, E. L.: Development of a Rigidized Surface Insulative Thermal Protection System for Shuttle Orbiter, Final Report. NASA CR-114973, Feb. 15, 1971.

4. Banas, R. P.; Kural, M. H.; Deruntz, J. A.; Burns, A. B.; Chinn, A. J.; Lambert, R.; Ritz, J. R.; Vanwest, B.; and Woneis, J. T.: Space Shuttle Thermal Protection System Development. Lockheed Report LMSC-D152738, Lockheed Missiles and Space Co., Space Systems Division, Jan. 17, 1972.

5. LI-1500 Heat Shield, Independent Development Program C548. Lockheed Missiles and Space Co. Report LMSC-D153908, Dec. 24, 1971.

6. Space Shuttle Concepts. Final Report for NAS8-26362, Lockheed Missiles and Space Co. Report LMSC-D153024, Mar. 15, 1972.

7. Development and Design Application of Rigidized Surface Insulation Thermal Protection Systems. Lockheed Missiles and Space Co. Report LMSC-D282673, Dec. 30, 1972.

8. A Proposal for Space Shuttle Orbiter High-Temperature Reusable Surface Insulation. Lockheed Missiles and Space Company Report LMSC-D336595, Mar. 14, 1973.

9. Forsberg, K.: Producing the High-Temperature Reusable Surface Insulation for the Thermal Protection System of the Space Shuttle. Presented at XIVe Congress International Aeronautique, Paris, France, June 1979.

10. Burns, A. Bruce and McCarter, C. R.: Manufacture of Reusable Surface Insulation (RSI) for the Space Shuttle Orbiter. Presented at the 1980 SAE Aerospace Congress & Exposition, Oct. 15, 1980.

11. Banas, R. P.; Dolton, T. A.; Housten, S. J.; and Wilson, R. G.: Lifting Entry Vehicle Thermal Protection Review. Proceedings of Thermodynamics and Thermophysics of Space Flight. Palo Alto, California, 1970, pp. 239-276.

12. Improvement of Reusable Surface Insulation Material. Final Report for NAS9-12137. Lockheed Missiles and Space Co. Report LMSC-D266204, Mar. 1, 1972.

13. Goldstein, H. E., Smith, Marnell and Leiser, Daniel: Silica Reusable Surface Insulation. United States Patent 3,952,083, Apr. 20, 1976.

14. Leiser, Daniel B., Goldstein, Howard, E. and Smith, Marnell: Fibrous Refractory Composite Insulation. United States Patent 4,148,962, Apr. 10, 1979 (FRCI-12).

15. Banas, R. P. and Cordia E. R.: Advanced High-Temperature Insulation Material For Reentry Heat Shield Applications. Presented at 4th Annual Conference On Composites and Advanced Materials, Ceramic - Metal Systems Division - American Ceramic Society, Jan. 1980.

16. Wheeler, W. H., Garofalini, S. H. and Beasley, R. M.: Development of an Unusual Coating System for the Space Shuttle Orbiters. Paper presented at American Ceramic Society Meeting in Cincinnati, Ohio, May 4, 1976.

17. Goldstein, H. E., Leiser, D. B. and Katvala, V. W.: Reaction Cured Glass and Glass Coatings. United States Patent No. 4,093,771, June 6, 1978.

18. Banas, R. P., Gzowski, E. R. and Larsen, W. T.: Processing Aspects of the Space Shuttle Orbiter's Ceramic Reusable Surface Insulation. Proceedings of the 7th Conference on Composites and Advanced Materials, Ceramic-Metal Systems Division - American Ceramic Society, Cocoa Beach, Fla.. Jan. 16-21, 1983.

19. Korb, L. J. and Clancy, H. M.: Shuttle Orbiter Thermal Protection System: A Material And Structural Overview. Presented at the 26th National Symposium Society for the Advancement of Materials and Processes Engineering, Paper No. STS 81-0219, Apr. 28-30 1981.

20. Dotts, R. L., Battley, H. H., Hughes, J. T. and Neuenschwander, W. E.: Space Shuttle Orbiter - Reusable Surface Insulation (RSI) Flight Performance. AIAA paper No. 82-0788-CP, May 1982.

21. Holmquist, G. R. and Cordia E. R.: Advances in Reusable Surface Insulation for Space Shuttle Application. MATERIALS 1980 - Proceedings of the 12th National SAMPE Technical Conference, Volume 12. Society for Advancement of Materials and Processes Engineering, Oct. 1980.

22. Banas, R. P. and Cunnington Jr., G. R.: Determination of Effective Thermal Conductivity for the Space Shuttle Orbiter's Reusable Surface Insulation (RSI). AIAA Paper No. 74-730, Thermophysics & Heat Transfer Conference, Boston, Ma., July 1974.

23. Williams, S. D. and Curry, D. M.: Nonlinear Least Squares - An Aid to Thermal Property Determination. NASA TM X-58092, June 1972.

24. Holmquist G. R., Cordia E. R. and Tomer, R. S.: Effects of Composition and Processing on Thermal Performance of a Rigidized Fibrous Ceramics Insulation Material. Ceramic Engineering and Science Proceedings - 5th Annual Conference on Composites and Advanced Materials, Volume 2, American Ceramic Society, July 1981.

25. Elgin, D. and Schirle, J.: Thermal Performance of a Fibrous Refractory Composite Insulation (FRCI). Presented at 5th Annual Conference on Composites and Advanced Materials. American Ceramic Society. Merritt Island, Florida. Jan. 18-22, 1981.

26. Tomer, R. S. and Cordia, E. R.: Development of an Improved Lightweight
    Insulation Material for the Space Shuttle Orbiter's Thermal
    Protection System, Ceramic Engineering and Science Proceedings – 6th
    Annual Conference on Composites and Advanced Materials, Volume 3,
    American Ceramic Society, pp. 601-611, Sept. 1982.

27. Tanabe, T. M. Izu, Y. D.: Influence of Particle Size Distribution on
    Ceramic Coating Characteristics. Presented at 32nd Pacific Boast
    Regional Meeting of American Ceramic Society, Seattle, Washington, Oct.
    24-26, 1979.

28. Dotts, R. L., Smith, J. A., and Tillian, D. J.: Space Shuttle Orbiter
    Reusable Surface Insulation Flight Results. Shuttle Performance:
    Lessons Learned, NASA CP-2283, Part 2, 1983, pp. 949-966.

29. Creedon, J. F., Banas, R. P. and Stern, P.: Results of Lockheed Missiles
    & Space Co. Studies of NASA/ARC Reaction Cured Glass Coating for the
    Space Shuttle Orbiter. Presented at American Ceramic Society's 29th
    Pacific Coast Meeting under Ceramic-Metal Systems Division, San Francisco,
    Ca.,Nov. 2, 1976.

30. Nakano, H. N. and Izu, Y. D.: Significance of Particle Size Distribution
    on Refractory Frit Sintering. Presented at 12th Annual Meeting of Fine
    Particle Society – Powder and Bulk Solids Conference, Rosemont, Illinois,
    May 12-14, 1981.

31. Fitchett, B. T.: Dimensional Control of Space Shuttle Tiles during
    Manufacture. Proceedings of the 7th Conference on Composites and
    Advanced Materials, Ceramic – Metal Systems Division, American Ceramic
    Society, Cocoa Beach, Fla.,Jan. 16-21, 1983.

32. Leiser, D. L., Smith, M., Stewart, D. and Goldstein, H.: Thermal and
    Mechanical Properties of Advanced High-Temperature Ceramic Composite
    Insulation. Proceedings of the 7th Conference on Composites and Advanced
    Materials, Ceramic – Metal Systems Division – American Ceramic Society,
    Cocoa Beach, Fla.,Jan. 16-21, 1983.

TABLE I.- MASTER DIMENSIONS REFINEMENTS

ORIGINAL METHOD	IMPROVED METHOD
• HAND CALCULATION OF TILE CORNER POINTS & PA STANDARDS.	• MD DEVELOPED SOFTWARE TO CALCULATE CORNER POINTS & PA STANDARDS.
• HAND CALCULATION OF BLANK SIZES.	• SOFTWARE DEVELOPED TO PROVIDE BLANK SIZES.
• APPROXIMATION OF COMPLEX SURFACES.	• CAD/CAM DEVELOPMENT & CHECK OF COMPLEX SURFACES.
• PRODUCT ASSURANCE POINTS FOR HARD-TO-INSPECT TILES.	• CAD/CAM DEVELOPED MYLARS TO INSPECT COMPLEX TILES.
• HAND CALCULATION OF POINTS TO CHECK ROCKWELL MYLARS.	• CAD/CAM CHECK OF ROCKWELL MYLARS.

TABLE II.- NC PROGRAMMING REFINEMENTS

ORIGINAL METHOD	IMPROVED METHOD
• CRUDE CUT PACKAGES TO PROGRAM TILE FAMILIES.	• OPTIMIZED CUT PACKAGES TO FACILITATE PROGRAMMING & REDUCE COMPUTER RUN TIME.
• ONE INCH DIAMETER CUTTER USED FOR MOST MACHINING.	• ABOUT 30 CUTTER GEOMETRIES WERE USED TO REDUCE MACHINE TIME
• NO COATING LINE ON COMPLEX TILES.	• COATING TERMINATOR LINE ADDED TO COMPLEX TILES REDUCED SCRAP.
• SEPARATE FIXTURES FOR DIFFERENT BLANK SIZES.	• SAME FIXTURE FOR ALL BLANKS REDUCED SCRAP.
• PLANAR SIDES PROGRAMMED AS PLANAR SURFACES.	• PLANAR SIDES PROGRAMMED AS CYLINDERS TO REDUCE CORNER SHRINKAGE.
• LIMITED KNOWLEDGE OF COMPLEX SURFACES.	• EXPERIENCE REDUCED PROGRAMMING & MACHINING TIME.
• LIMITED USE OF CAD/CAM SYSTEM TO PROGRAM TILES.	• CAD/CAM SYSTEM SHOULD REDUCE TOOL TRIES.
• NO CHECK FOR REFERENCE POINT	• HOMING DEVICES INSTALLED ON N/C MILLS ASSURE PROPER REFERENCE POINT

TABLE III.- A SUMMARY OF LESSONS LEARNED

1. FIBER PREPARATION METHODS
   o A 20 blend of Manville silica fibers provides better material uniformity.
   o Use of a hydro-cyclone removes unwanted glass shot from the silica fiber lots.
   o Dual stage agitation of Nextel fibers eliminates fiber clumping in FRCI-12 billets.

2. BILLET CASTING AND SINTERING METHODS
   o Precursor silica fiber lot compact tests allow better prediction of production billet sintering requirements.
   o Elimination of an intermediate billet sintering cycle for LI-900 was successfully accomplished.
   o Implementation of void reduction and vacuum assisted casting procedures minimize voids in LI-900 billets.
   o Implementation of six-side heating for LI-900 and FRCI-12 during the sintering cycle improved the strength distribution within billets.
   o Vacuum degassing of slurries for LI-2200 and FRCI-12 prior to casting eliminates voids in the billets.
   o The high dry-density concept for FRCI-12 and FRCI-8 was successfully used to tailor the apparent thermal conductivity.

3. ENGINEERING DATA AND NUMERICAL CONTROL PROGRAMMING REFINEMENTS
   o Experience with the data has led to many refinements in the tile part programs that have made tile fabrication more efficient.
   o Addition of various software tools reduced the requirement to hand calculate various master dimension surfaces.
   o The use of interactive graphics methods for tile design and NC part programming would reduce the number of tool tries and improve efficiency.
   o Use of CADAM to generate mylars for use as tile dimensional inspection tools eliminates the need to supply computerized tile dimensional data for hard-to-measure tiles.

4. TILE FABRICATION
   o The addition of homing devices on the NC mills provided assurance that tiles were being cut correctly.
   o Design changes to the cutting tools improved the consistency of the tile-to-tile gaps in the delivered tiles.
   o An understanding of fiber shrinkage characteristics for LI-900, LI-2200 and FRCI-12 led to the implementation of tile machining offsets that resulted in tiles that meet the dimensional requirements.
   o Implementation of letter offsets on the IBM tile travel cards with appropriate changes in the NC hardware and software provided an efficient method to modify tile machining offsets.
   o Recognition of the anisotropic shrinkage characteristics of LI-900, LI-2200 and FRCI-12 further improved the tile dimensional yield.

5. COATINGS AND THEIR APPLICATION
   o Use of a universal patch compound prior to coating spraying eliminated irregular surfaces and craters in the glazed coatings.
   o Man is a more efficient tile sprayer than robots.
   o Control of particle size for glass frits and coating slurries eliminated most coating anomalies.
   o Real time process control of coating weight in the manufacturing area eliminated tiles scrapped for coating weight.
   o The propensity for cracks in the Class 1 coating was resolved by increasing the maximum coating weight to 0.17 $lb/_{ft}2$, making it consistent with the Class 2 coating.

6. TILE MEASURMEENT AND SHIMMING
   o Introduction of the "maxi-measure" and Cordax apparatus yielded accurate dimensional data.
   o The "Load-and-Go" concept minimized tiles being scrapped for planform dimensional anomalies, while preserving dimensional control at the array level.
   o Introduction of a second glazing operation, to shrink tiles, eliminated tiles scrapped for an oversize condition.
   o Use of mylars for complex or hard-to-measure tiles for Challenger greatly improved the determination of acceptability compared to the "ship-and-fit" criterion used for Columbia's complex tiles.

- **TOTAL RSI CERAMIC TILES - 30,812 (LMSC 23,400)**
- **REINFORCED CARBON/CARBON (RCC) (44 PANELS/NOSE CAP)**
- **FELT REUSABLE SURFACE INSULATION (FRSI) (3,581 FT^2)**

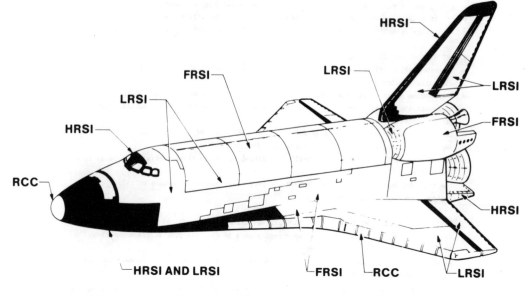

Figure 1.- TPS locations on Columbia (OV-102).

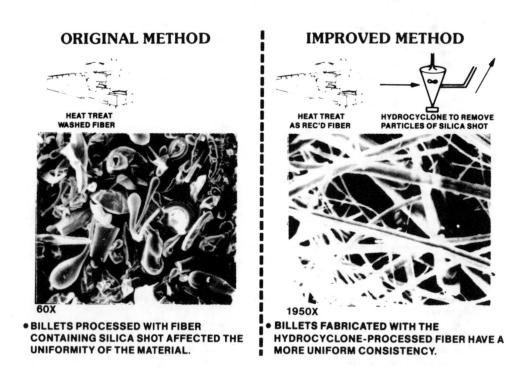

Figure 2.- Pretreatment of silica fibers for LI-900.

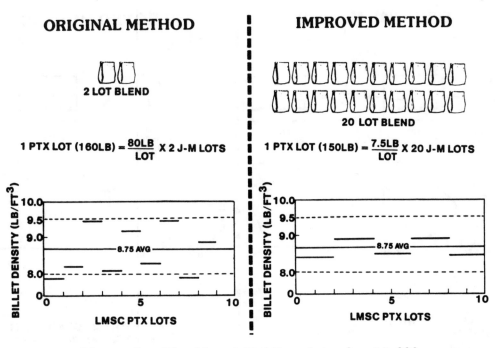

Figure 3.– Blending J-M fiber lots for LI-900.

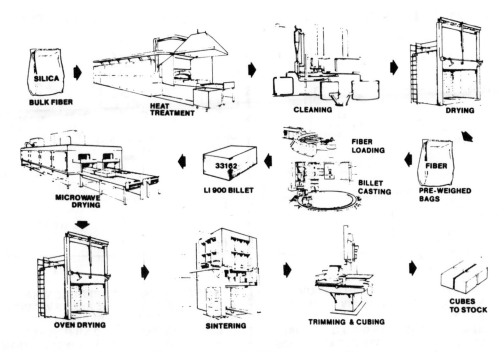

Figure 4.– LI-900 process flow.

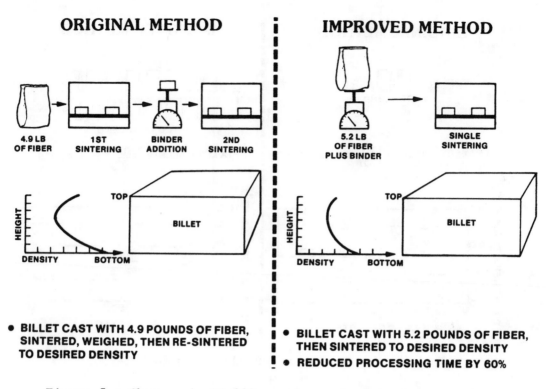

ORIGINAL METHOD

4.9 LB OF FIBER → 1ST SINTERING → BINDER ADDITION → 2ND SINTERING

HEIGHT / DENSITY / TOP / BOTTOM / BILLET

- BILLET CAST WITH 4.9 POUNDS OF FIBER, SINTERED, WEIGHED, THEN RE-SINTERED TO DESIRED DENSITY

IMPROVED METHOD

5.2 LB OF FIBER PLUS BINDER → SINGLE SINTERING

HEIGHT / DENSITY / TOP / BOTTOM / BILLET

- BILLET CAST WITH 5.2 POUNDS OF FIBER, THEN SINTERED TO DESIRED DENSITY
- REDUCED PROCESSING TIME BY 60%

Figure 5.- Changes in LI-900 casting and sintering procedures.

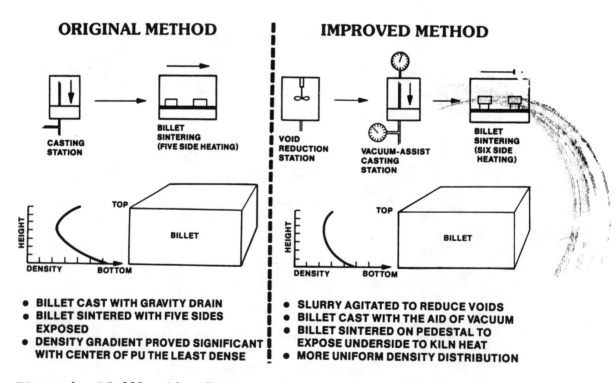

ORIGINAL METHOD

CASTING STATION → BILLET SINTERING (FIVE SIDE HEATING)

HEIGHT / DENSITY / TOP / BOTTOM / BILLET

- BILLET CAST WITH GRAVITY DRAIN
- BILLET SINTERED WITH FIVE SIDES EXPOSED
- DENSITY GRADIENT PROVED SIGNIFICANT WITH CENTER OF PU THE LEAST DENSE

IMPROVED METHOD

VOID REDUCTION STATION → VACUUM-ASSIST CASTING STATION → BILLET SINTERING (SIX SIDE HEATING)

HEIGHT / DENSITY / TOP / BOTTOM / BILLET

- SLURRY AGITATED TO REDUCE VOIDS
- BILLET CAST WITH THE AID OF VACUUM
- BILLET SINTERED ON PEDESTAL TO EXPOSE UNDERSIDE TO KILN HEAT
- MORE UNIFORM DENSITY DISTRIBUTION

Figure 6.- LI-900 void reduction and vacuum assisted casting methods.

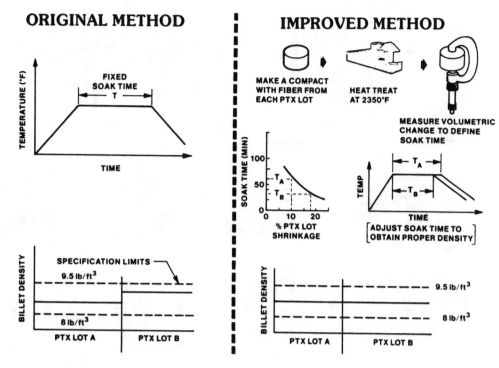

ORIGINAL METHOD

IMPROVED METHOD

MAKE A COMPACT WITH FIBER FROM EACH PTX LOT

HEAT TREAT AT 2350°F

MEASURE VOLUMETRIC CHANGE TO DEFINE SOAK TIME

ADJUST SOAK TIME TO OBTAIN PROPER DENSITY

Figure 7.- Silica fiber compact test for LI-900.

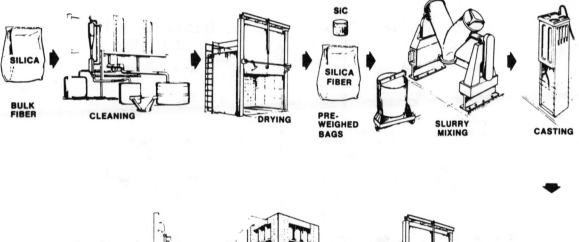

Figure 8.- LI-2200 process flow.

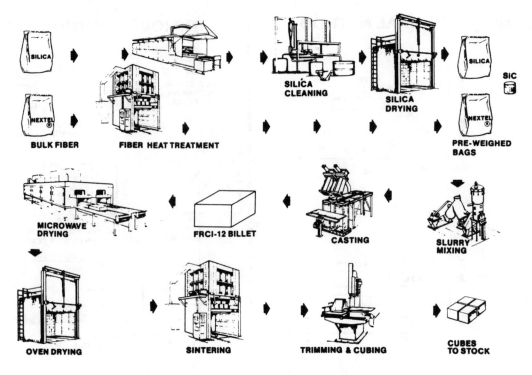

Figure 9.- FRCI-12 process flow.

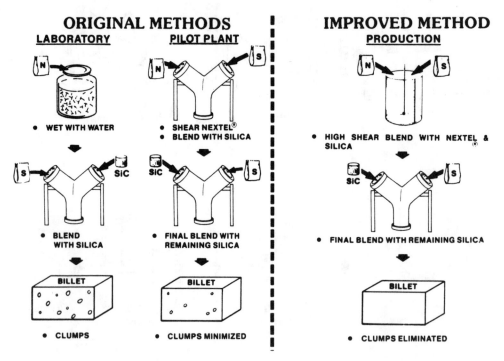

Figure 10.- Nextel dispersion methods for FRCI-12.

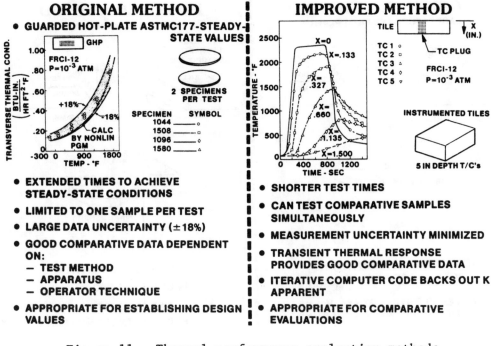

ORIGINAL METHOD

- **GUARDED HOT-PLATE ASTMC177-STEADY-STATE VALUES**

- **EXTENDED TIMES TO ACHIEVE STEADY-STATE CONDITIONS**
- **LIMITED TO ONE SAMPLE PER TEST**
- **LARGE DATA UNCERTAINTY (±18%)**
- **GOOD COMPARATIVE DATA DEPENDENT ON:**
  - **TEST METHOD**
  - **APPARATUS**
  - **OPERATOR TECHNIQUE**
- **APPROPRIATE FOR ESTABLISHING DESIGN VALUES**

IMPROVED METHOD

- **SHORTER TEST TIMES**
- **CAN TEST COMPARATIVE SAMPLES SIMULTANEOUSLY**
- **MEASUREMENT UNCERTAINTY MINIMIZED**
- **TRANSIENT THERMAL RESPONSE PROVIDES GOOD COMPARATIVE DATA**
- **ITERATIVE COMPUTER CODE BACKS OUT K APPARENT**
- **APPROPRIATE FOR COMPARATIVE EVALUATIONS**

Figure 11.- Thermal performance evaluation methods.

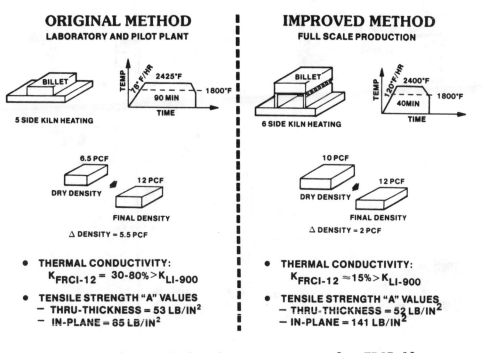

ORIGINAL METHOD
LABORATORY AND PILOT PLANT

5 SIDE KILN HEATING

Δ DENSITY = 5.5 PCF

- **THERMAL CONDUCTIVITY:**
  $K_{FRCI-12} = 30-80\% > K_{LI-900}$
- **TENSILE STRENGTH "A" VALUES**
  - THRU-THICKNESS = 53 LB/IN^2
  - IN-PLANE = 85 LB/IN^2

IMPROVED METHOD
FULL SCALE PRODUCTION

6 SIDE KILN HEATING

Δ DENSITY = 2 PCF

- **THERMAL CONDUCTIVITY:**
  $K_{FRCI-12} \approx 15\% > K_{LI-900}$
- **TENSILE STRENGTH "A" VALUES**
  - THRU-THICKNESS = 52 LB/IN^2
  - IN-PLANE = 141 LB/IN^2

Figure 12.- High dry density concept for FRCI-12.

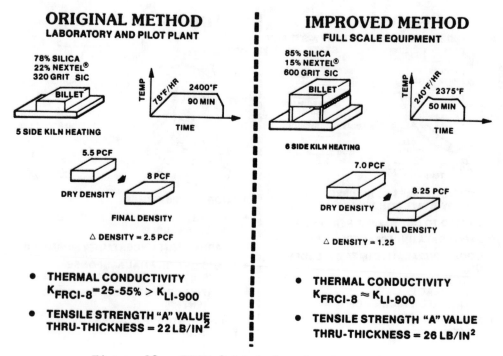

ORIGINAL METHOD
LABORATORY AND PILOT PLANT

78% SILICA
22% NEXTEL®
320 GRIT SIC

BILLET

5 SIDE KILN HEATING

TEMP — 78°F/HR — 2400°F — 90 MIN — TIME

5.5 PCF
DRY DENSITY

8 PCF
FINAL DENSITY

△ DENSITY = 2.5 PCF

- THERMAL CONDUCTIVITY
  $K_{FRCI-8} = 25\text{-}55\% > K_{LI\text{-}900}$

- TENSILE STRENGTH "A" VALUE
  THRU-THICKNESS = 22 LB/IN^2

IMPROVED METHOD
FULL SCALE EQUIPMENT

85% SILICA
15% NEXTEL®
600 GRIT SIC

BILLET

6 SIDE KILN HEATING

TEMP — 240°F/HR — 2375°F — 50 MIN — TIME

7.0 PCF
DRY DENSITY

8.25 PCF
FINAL DENSITY

△ DENSITY = 1.25

- THERMAL CONDUCTIVITY
  $K_{FRCI-8} \approx K_{LI\text{-}900}$

- TENSILE STRENGTH "A" VALUE
  THRU-THICKNESS = 26 LB/IN^2

Figure 13.- FRCI-8 high dry density concept.

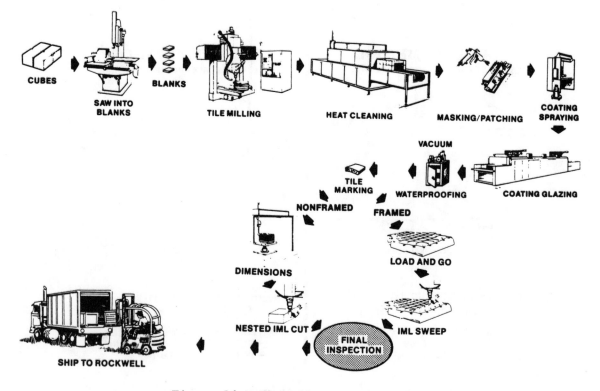

CUBES → SAW INTO BLANKS → BLANKS → TILE MILLING → HEAT CLEANING → MASKING/PATCHING → COATING SPRAYING

COATING GLAZING

VACUUM WATERPROOFING

TILE MARKING

NONFRAMED

FRAMED

LOAD AND GO

DIMENSIONS

NESTED IML CUT

FINAL INSPECTION

IML SWEEP

SHIP TO ROCKWELL

Figure 14.- The tile process flow.

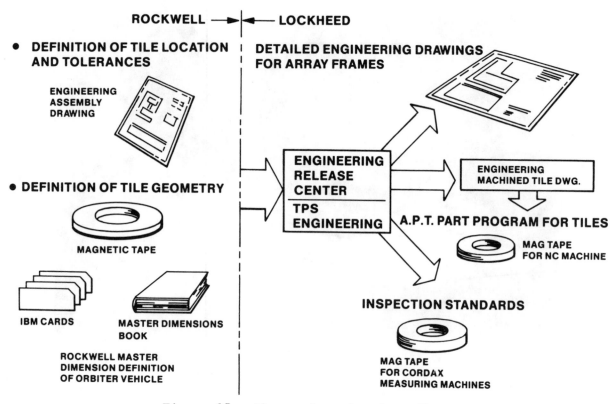

Figure 15.- The engineering data flow.

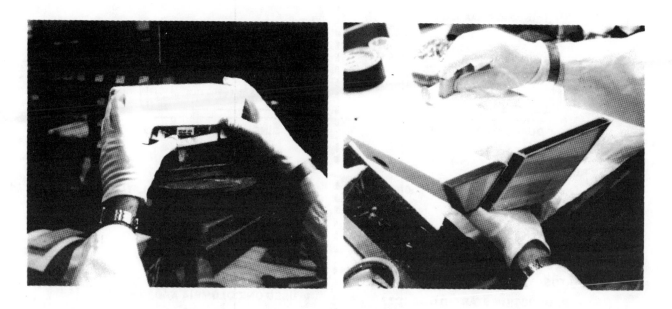

APPLYING MASKING TEMPLATE

PATCHING VOIDS

Figure 16.- Examples of complex tiles without coating terminator lines.

Figure 17.- An interactive graphics method showing a Shuttle infrared leeside temperature sensing tile.

ORIGINAL COATING

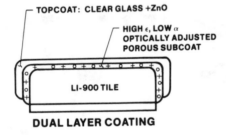

TOPCOAT: CLEAR GLASS +ZnO

HIGH ϵ, LOW α OPTICALLY ADJUSTED POROUS SUBCOAT

LI-900 TILE

DUAL LAYER COATING

REQUIREMENTS:

- $0.2 < \alpha/\epsilon < 0.4 - (135°F \text{ to } + 250°F)$
- $\epsilon \geq 0.8$ AT 1200°F
- WEIGHT ≤ 0.09 lb/ft^2
- CRACK FREE AND WATER IMPERVIOUS

STATUS:

- PRODUCTION START IN OCT. 1977
- THIN COATING CRACKED EASILY

IMPROVED COATING

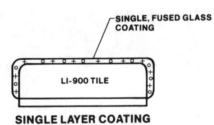

SINGLE, FUSED GLASS COATING

LI-900 TILE

SINGLE LAYER COATING

ADVANTAGES:

- ALL COMPONENTS IN ONE LAYER
- MINIMIZED WATER IMPERVIOUSNESS PROBLEM
- LESS RESIDUAL STRAIN

STATUS:

- USED ON COLUMBIA AT 0.12 lb/ft^2
- USED ON CHALLENGER AT 0.17 lb/ft^2

Figure 18.- Class 1 coating optimization.

ORIGINAL METHOD

- **COATING ANOMALIES**

"MUD CRACKING"   **"ONION SKINNING"**

STATUS:

- **NO CLASS 1 FRIT SIZE CONTROL OTHER THAN 90% THROUGH 325 MESH AND A 50% POINT REQUIREMENT**
- **USE OF SLURRY IMMEDIATELY AFTER PREPARATION ELIMINATED VISCOSITY VARIATIONS**
- **NO CLASS 2 FRIT SIZE CONTROL OTHER THAN A 50% POINT REQUIREMENT**

IMPROVED METHOD

- **DEFINED FULL PARTICLE SIZE DISTRIBUTION FOR CL.1/CL. 2 FRITS AND SLURRIES**

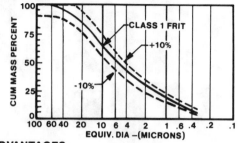

ADVANTAGES:

- **SLURRY VISCOSITY CONTROL IS IMPROVED**
- **SPRAYING CHARACTERISTICS ARE MORE UNIFORM**
- **COATING CRACKS DURING GLAZING ARE ELIMINATED**
- **FUSION OF CLASS 1 AND CLASS 2 COATINGS IS ASSURED**

Figure 19.- Effect of glass frit particle size effects on glazed coatings.

ORIGINAL COATING

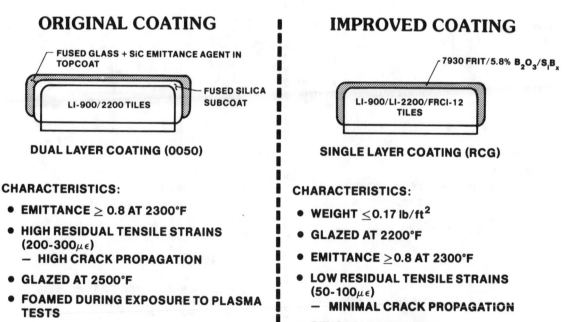

— FUSED GLASS + SiC EMITTANCE AGENT IN TOPCOAT

LI-900/2200 TILES

— FUSED SILICA SUBCOAT

DUAL LAYER COATING (0050)

CHARACTERISTICS:

- **EMITTANCE ≥ 0.8 AT 2300°F**
- **HIGH RESIDUAL TENSILE STRAINS (200-300$\mu\epsilon$)**
  - **HIGH CRACK PROPAGATION**
- **GLAZED AT 2500°F**
- **FOAMED DURING EXPOSURE TO PLASMA TESTS**

IMPROVED COATING

— 7930 FRIT/5.8% B_2O_3/S_iB_x

LI-900/LI-2200/FRCI-12 TILES

SINGLE LAYER COATING (RCG)

CHARACTERISTICS:

- **WEIGHT ≤0.17 lb/ft^2**
- **GLAZED AT 2200°F**
- **EMITTANCE ≥0.8 AT 2300°F**
- **LOW RESIDUAL TENSILE STRAINS (50-100$\mu\epsilon$)**
  - **MINIMAL CRACK PROPAGATION**
- **REUSABLE TO 2300°F**
- **GOOD FOR SINGLE EXPOSURE TO 2700°F**

Figure 20.- Class 2 coating optimization.

ORIGINAL METHOD

IMPROVED METHOD

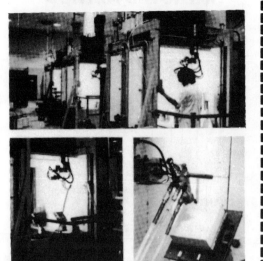

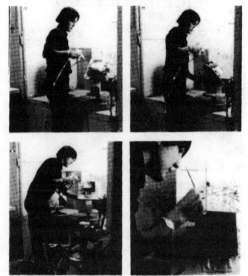

- ROBOTS FOR SIMPLE TITLES (5 COATED SURFACES)
- MAN FOR COMPLEX TITLES (UP TO 20 SURFACES)

- MAN IS THE MOST FLEXIBLE ROBOT FOR ALL TILES

Figure 21.- Methods of spraying class 1 and class 2 tiles.

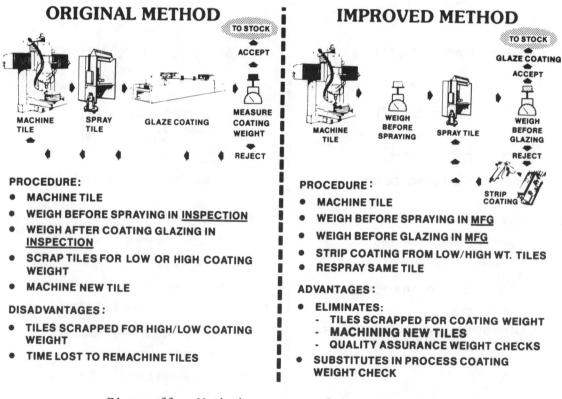

ORIGINAL METHOD

MACHINE TILE → SPRAY TILE → GLAZE COATING → MEASURE COATING WEIGHT

TO STOCK — ACCEPT / REJECT

PROCEDURE:
- MACHINE TILE
- WEIGH BEFORE SPRAYING IN <u>INSPECTION</u>
- WEIGH AFTER COATING GLAZING IN <u>INSPECTION</u>
- SCRAP TILES FOR LOW OR HIGH COATING WEIGHT
- MACHINE NEW TILE

DISADVANTAGES:
- TILES SCRAPPED FOR HIGH/LOW COATING WEIGHT
- TIME LOST TO REMACHINE TILES

IMPROVED METHOD

MACHINE TILE → WEIGH BEFORE SPRAYING → SPRAY TILE → WEIGH BEFORE GLAZING

TO STOCK — GLAZE COATING — ACCEPT / REJECT — STRIP COATING

PROCEDURE:
- MACHINE TILE
- WEIGH BEFORE SPRAYING IN <u>MFG</u>
- WEIGH BEFORE GLAZING IN <u>MFG</u>
- STRIP COATING FROM LOW/HIGH WT. TILES
- RESPRAY SAME TILE

ADVANTAGES:
- ELIMINATES:
  - TILES SCRAPPED FOR COATING WEIGHT
  - MACHINING NEW TILES
  - QUALITY ASSURANCE WEIGHT CHECKS
- SUBSTITUTES IN PROCESS COATING WEIGHT CHECK

Figure 22.- Methods to control coating weight.

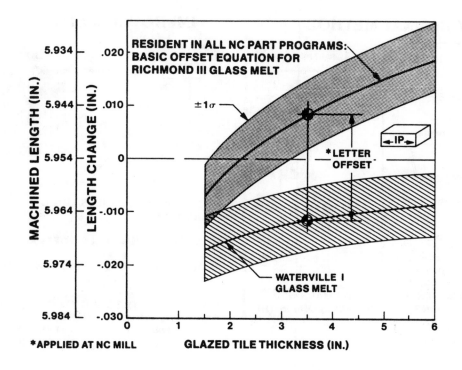

Figure 23.- In-plane machining offsets for LI-900.

- ALL ENGINEERING DEFINITION IS BASED ON COMPUTER (MASTER DIMENSION) DATA
- EACH TILE HAS UNIQUE PART PROGRAM FOR NC MACHINING
- ALL MANUFACTURING AND INSPECTION OPERATIONS ARE CONTROLLED BY ONE IBM CARD PER TILE

- DIMENSIONAL REQUIREMENT ± 0.016-INCH (LENGTH & WIDTH) AND ± 0.010-INCH (THICKNESS) FOR MOST TILES
- EACH TILE REQUIRES COMPENSATION FOR MATERIAL SHRINKAGE DURING COATING GLAZING
- SHRINKAGE VARIES WITH
  - GLASS MELT
  - TILE THICKNESS
  - TILE PLANFORM

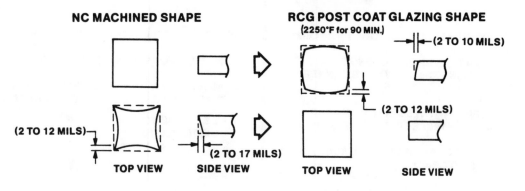

Figure 24.- Tile fabrication requirements.

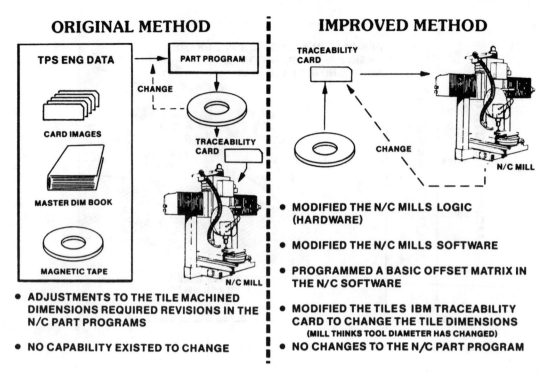

ORIGINAL METHOD

TPS ENG DATA

CARD IMAGES

MASTER DIM BOOK

MAGNETIC TAPE

PART PROGRAM

CHANGE

TRACEABILITY CARD

N/C MILL

- ADJUSTMENTS TO THE TILE MACHINED DIMENSIONS REQUIRED REVISIONS IN THE N/C PART PROGRAMS

- NO CAPABILITY EXISTED TO CHANGE

IMPROVED METHOD

TRACEABILITY CARD

CHANGE

N/C MILL

- MODIFIED THE N/C MILLS LOGIC (HARDWARE)

- MODIFIED THE N/C MILLS SOFTWARE

- PROGRAMMED A BASIC OFFSET MATRIX IN THE N/C SOFTWARE

- MODIFIED THE TILES IBM TRACEABILITY CARD TO CHANGE THE TILE DIMENSIONS (MILL THINKS TOOL DIAMETER HAS CHANGED)

- NO CHANGES TO THE N/C PART PROGRAM

Figure 25.- Implementation aspects of tile machining letter offsets.

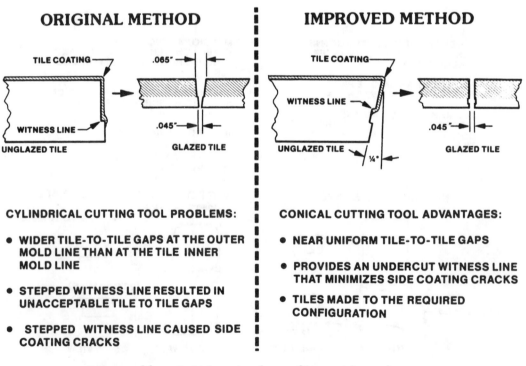

ORIGINAL METHOD

TILE COATING

WITNESS LINE

UNGLAZED TILE

.065"

.045"

GLAZED TILE

CYLINDRICAL CUTTING TOOL PROBLEMS:

- WIDER TILE-TO-TILE GAPS AT THE OUTER MOLD LINE THAN AT THE TILE INNER MOLD LINE

- STEPPED WITNESS LINE RESULTED IN UNACCEPTABLE TILE TO TILE GAPS

- STEPPED WITNESS LINE CAUSED SIDE COATING CRACKS

IMPROVED METHOD

TILE COATING

WITNESS LINE

UNGLAZED TILE

¼°

.045"

GLAZED TILE

CONICAL CUTTING TOOL ADVANTAGES:

- NEAR UNIFORM TILE-TO-TILE GAPS

- PROVIDES AN UNDERCUT WITNESS LINE THAT MINIMIZES SIDE COATING CRACKS

- TILES MADE TO THE REQUIRED CONFIGURATION

Figure 26.- Cutting tool configuration changes.

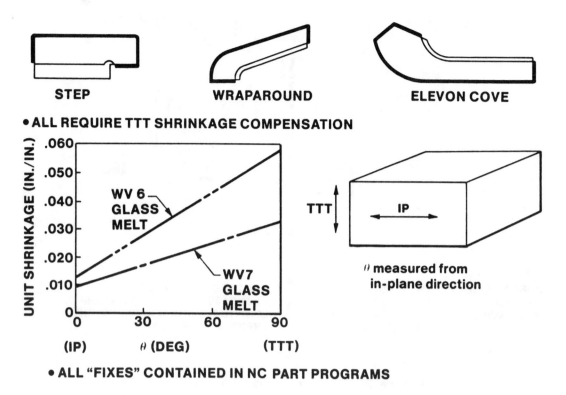

STEP   WRAPAROUND   ELEVON COVE

● ALL REQUIRE TTT SHRINKAGE COMPENSATION

θ measured from in-plane direction

● ALL "FIXES" CONTAINED IN NC PART PROGRAMS

Figure 27.- LI-900 tile shrinkage in the through-the-thickness direction.

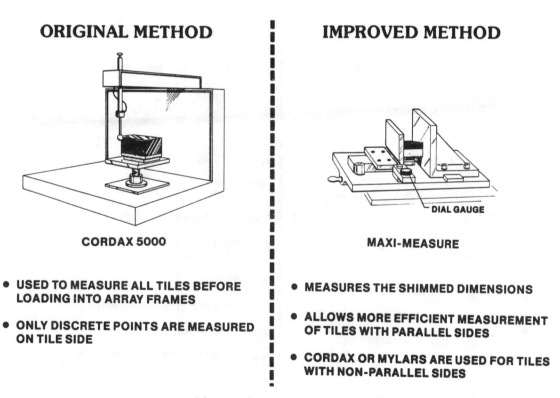

ORIGINAL METHOD

CORDAX 5000

● USED TO MEASURE ALL TILES BEFORE LOADING INTO ARRAY FRAMES

● ONLY DISCRETE POINTS ARE MEASURED ON TILE SIDE

IMPROVED METHOD

DIAL GAUGE

MAXI-MEASURE

● MEASURES THE SHIMMED DIMENSIONS

● ALLOWS MORE EFFICIENT MEASUREMENT OF TILES WITH PARALLEL SIDES

● CORDAX OR MYLARS ARE USED FOR TILES WITH NON-PARALLEL SIDES

Figure 28.- Tile measurement methods.

ORIGINAL METHOD

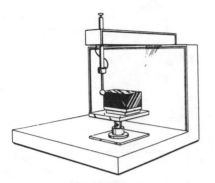

- CORDAX 5000 MEASURING DEVICE WAS USED TO MEASURE ALL TILES BEFORE THEY WERE LOADED INTO ARRAY FRAMES. THIS MADE IT NECESSARY TO GENERATE STANDARDS THAT DEFINED THE CONFIGURATION OF EACH TILE.

IMPROVED METHOD

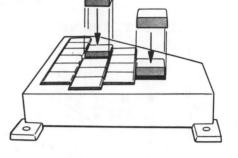

- ALL TILES ARE LOADED INTO PREMEASURED ARRAY FRAMES. IF TILES LOAD TO SPECIFIED GAPS, THE ENTIRE FRAME IS ACCEPTED FOR SHIPMENT

Figure 29.- The load and go concept.

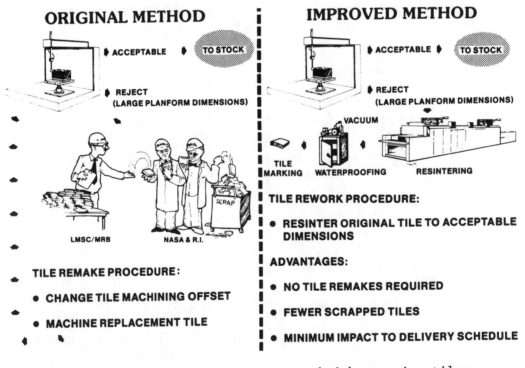

ORIGINAL METHOD

ACCEPTABLE → TO STOCK

REJECT (LARGE PLANFORM DIMENSIONS)

LMSC/MRB   NASA & R.I.

SCRAP

TILE REMAKE PROCEDURE:

- CHANGE TILE MACHINING OFFSET
- MACHINE REPLACEMENT TILE

IMPROVED METHOD

ACCEPTABLE → TO STOCK

REJECT (LARGE PLANFORM DIMENSIONS)

VACUUM

TILE MARKING   WATERPROOFING   RESINTERING

TILE REWORK PROCEDURE:

- RESINTER ORIGINAL TILE TO ACCEPTABLE DIMENSIONS

ADVANTAGES:

- NO TILE REMAKES REQUIRED
- FEWER SCRAPPED TILES
- MINIMUM IMPACT TO DELIVERY SCHEDULE

Figure 30.- A second sintering to shrink oversize tiles.

DENSITY (lb/ft³)	LI-900	LI-2200	FRCI-12
	8.0-9.5	20-24	11.9-13.5
TENSILE STRENGTH* (lb/in²) THRU-THE-THICKNESS IN-PLANE	24 67	73 180	81 257
COMPRESSIVE STRENGTH* (lb/in²) THRU-THE-THICKNESS IN-PLANE	28 70	130 230	132 265
THERMAL EXPANSION* (in/in - °F) THRU-THE-THICKNESS IN-PLANE	4×10^{-7} 4×10^{-7}	4×10^{-7} 4×10^{-7}	7×10^{-7} 7×10^{-7}
APPARENT THERMAL CONDUCTIVITY* (BTU-in/ft² hr - °F) THRU-THE-THICKNESS 70°F @ 10^{-4} ATM 1000°F @ 10^{-4} ATM IN-PLANE 70°F @ 1 ATM 1000°F @ 1 ATM	0.10 0.28 0.44 1.08	0.22 0.41 0.73 1.25	0.13 0.34 0.53 1.13
SPECIFIC HEAT* (BTU/lb - °F)	0.17	0.17	0.17

*AVERAGE VALUE

Figure 31.- Typical physical properties of LI-900, LI-2200 and FRCI-12.

811093

Commercial Filament Wound Pressure Vessels for Military and Aerospace Applications

Edgar E. Morris
Structural Composites Industries
Pomona, CA

IN THE 1960s, LIGHT-WEIGHT HIGH-strength pressure vessels were needed for space, missile, and military aircraft and underwater systems. Programs were sponsored by NASA-Lewis Research Center and the Air Force Materials Laboratory to design, develop, fabricate and evaluate composite tanks made by overwrapping thin metal liners with continuous high strength fibers embedded in resins based on technology developed for solid rocket motor cases. The resulting pressure vessels were 20 to 50 percent lighter than homogeneous metal vessels and offered other advantages, including enhanced safety with leakage mode-of-final-failure instead of catastrophic rupture.

Then, in the early 1970s, NASA was faced with the demand to help commercialize government-sponsored technology. In response, the Johnson Spacecraft Center awarded contracts for the development, pilot manufacture and service test of an improved light-weight commercial fireman's breathing system. Under this program, SCI fabricated aluminum cylinders fully reinforced by S-glass filaments. These 4000 psig operating pressure composite tanks were used in field tests by the Fire Departments of New York City, Los Angeles and Houston with excellent results. Similar fireman's breathing system composite cylinders with 2216 psig operating pressure have been commercially produced by SCI for Mine Safety Appliances Company since 1976, and the 50,000th unit was recently delivered.

In 1974, the Boeing Commercial

ABSTRACT

Light weight, low cost, composite pressure vessels, fabricated by over-wrapping thin metal liners with continuous high strength fibers embedded in resins, have been developed as a result of transferring aerospace technology to commercial products. Because the resulting filament wound tanks offer up to 50% reduction in weight and other advantages compared to homogeneous metal vessels, they have been successful in numerous commercial applications. With their advanced technology but lower cost, composite tanks are perfect for military and aerospace applications - as evidenced by the recent selection of commercial filament wound cylinders for the Manned Maneuvering Unit thruster for Space Shuttle astronauts.

Airplane Company decided to use composite reinforced cylinders for escape slide inflation to meet weight objectives on their long range 747 Aircraft. SCI developed a Kevlar/aluminum tank for this application and in 1976 received the first Department of Transportation authorization for routine commercial use of a composite reinforced cylinder.

Since then many other filament wound tanks have been developed for commercial use. So successful has this technology transfer from government to commercial applications been that now commercial composite pressure vessels are being chosen for military and aerospace applications. For example, NASA selected commercial filament wound tanks, developed by SCI in 1979, for the Manned Maneuvering Unit (MMU) of Space Shuttle.

This paper describes composite pressure vessels and explains their development, fabrication, advantages and uses.

DESCRIPTION

Composite pressure vessels are usually cylinders or spheres. Sizes of currently available composite cylinders range in volume from approximately 45 to over 10,000 cubic inches with diameters of 3 to over 20 inches. Operating pressures are normally from 2000 to 4500 psig; burst pressures must be three times operating pressure to comply with government regulations for commercial composite cylinders. Aerospace spheres produced have ranged in volume from 32 to over 27,000 cubic inches with diameters from 4 to over 38 inches and operating pressures of 2000 to 4500 psig with burst pressure 1.5 to 3 times operating pressure.

Presently, commercial tanks are cylinders made from thin metal liners overwrapped with continuous high strength fibers embedded in resins. The liner's purpose is to prevent leakage and provide a boss for the valve. However, it also serves as a filament winding mandrel during production and shares the pressure load during service. The usual liner material is seamless aluminum, but steel and other materials are also utilized. A seamless liner without welds or joints is desirable to minimize potential failures from fatigue or mishandling; however, welded liners are used for special applications. E-glass (240,000 psi minimum tensile strength), S-glass (350,000 psi minimum tensile strength) and Kevlar-49 (340,000 psi minimum tensile strength) fibers are usually used for overwrapping; however, the cost of Kevlar ($\sim$ \$10/lb) often limits its use to aircraft, military and aerospace applications. Although graphite fibers are a possibility, there are no known graphite units in commercial production - largely because of the high cost of graphite fibers without potential for significant additional weight saving. The glass or Kevlar fibers, in a resin system that is usually epoxy with either amine or anhydride curing, are filament wound over the liner.

Some future commercial tanks will certainly be composite spheres. They can directly replace metal spheres in existing pneumatic systems or be designed for new systems where the available envelope dictates a spherical shape.

DEVELOPMENT

The origin of commercial composite pressure vessels lies in the aerospace/defense industry fabrication of filament wound solid rocket motor cases. These single-use devices have to contain propellant combustion gases at high pressure (1000-1500 psi) and temperature (3000-5000°F) for short periods ($\sim$ 150 sec) and be as light weight as possible. Their liners are usually a rubber insulation material. The filament wound motor cases have polar bosses at both ends to accommodate an ignitor and a nozzle. The transfer of this technology to commercial pressure vessel products required the development of metal liners for long term storage, composite wrappings to carry most of the structural load and environmentally-resistant resin systems for the wrapped fibers to survive a variety of conditions for approximately 20 years of repeated use. To reduce cost and weight, the number of bosses was reduced from two to one for most applications; the single boss was converted to an end fitting for a valve or regulator and made part of the liner. Thus, there is no metal-to-composite joint. Since the pressure vessel loading on the composite structure is almost entirely tensile (without bending or shear), the composite cylinder is an ideal use of high strength fibers.

However, the composite pressure vessel had not been proven in commercial applications. High pressure gases (usually at 2000-4500 psi) are considered dangerous commodities, and their transportation in interstate commerce makes their containers subject to the regulations of the Department of Transportation. Steel cylinder regulations (called specifications), which generally reflect the codification of the existing practice when they were written many years ago, are well documented, but there are no specifications for composite cylinders. Thus, their uses are authorized by "Exemptions" to the specifications. To demonstrate that the safety of a composite cylinder is equivalent to a steel cylinder, extensive qualification tests are required for each type of composite cylinder. Since receiving the first such exemption in 1976, SCI has obtained others as needed for new pressure vessel models.

FABRICATION

LINER - Aluminum liners for composite cylinders can be fabricated by:
- o Impact Extrusion
  This is a two step process. First is an impact to make an open cup, and second is an impact on the open end of the cup to close it and form the boss.
- o Deep Draw and Spin
  This is a multi-step operation, as shown in Fig. 1, with intermediate anneals. The final step is spin closing without a mandrel. Although more expensive than impact extrusion, this process permits thin wall liners with close tolerances on thicknesses.

Resulting aluminum liners (6061 or 6351) are heat treated to the T6 condition and then machined to shape the boss and provide threads for the valve.

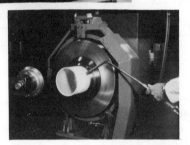

Fig. 1 - Deep draw and spin liner fabrication

OVERWRAP - The liner is completely filament wound with fiberglass or Kevlar in epoxy resin using interspersed helical and hoop patterns as illustrated in Figs. 2 and 3.

Helical Winding   Circumferential (Hoop) Winding

Fig. 2 - Winding patterns

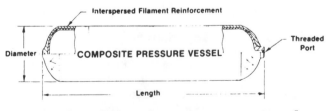

Fig. 3 - Composite pressure vessel

A typical two spindle helical winding machine is shown in Fig. 4.

Fig. 4 - Typical two spindle helical filament winding machine

After overwrapping, a polyurethane coating of the desired color is applied to the cylinder (Fig. 5) and it is cured (Fig. 6) in an oven at 300°F. Completed cylinders are shown in Fig. 7.

LABEL - The label on a composite cylinder is often overwrapped in place with clear fiberglass hoop wrap. The cylinder serial number is stamped on the side or end of the boss to provide additional positive identification. Retest dates can be stamped on the exposed neck of the cylinder.

TESTING - Some of the tests required by the DOT to assure that

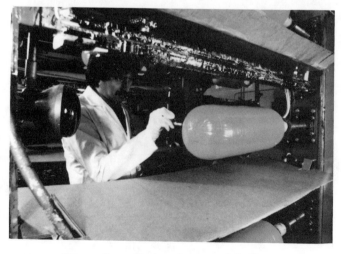

Fig. 5 - Painting cylinder

Fig. 6 - Oven curing cylinder

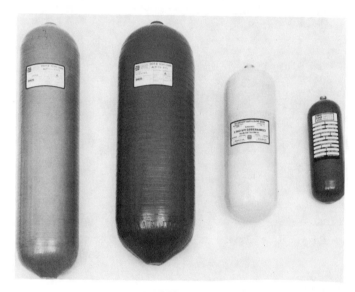

Fig. 7 - Typical fully wrapped cylinders

the composite cylinder is as safe as the all metal cylinder for transportation are as follows: acceptance hydrotest including test pressure application (5/3's operating pressure) and measurement of temporary and permanent expansion (Fig. 8); and product qualification tests including pressure and temperature cycle (Fig. 9), burst (Fig. 10), gunfire (Fig. 11), bonfire (Fig. 12), flaw growth resistance (Fig. 13), and impact (Fig. 14).

An independent inspection system ensures that production units are made in the same manner as the test units. There is also non-destructive inspection of all units (Fig. 15) plus destructive testing on 1 or 2

Fig. 8 - Vessel hydrotest

Fig. 11 - Gunfire test showing no fragmentation of tank

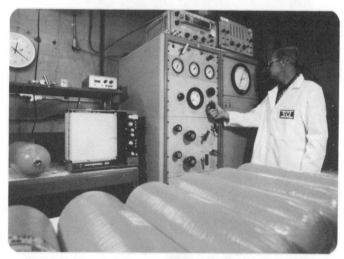

Fig. 9 - Pressure cycling of vessel

Fig. 12 - Bonfire test showing no explosion

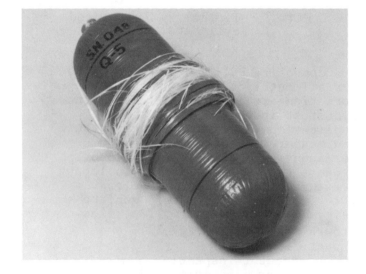

Fig. 10 - Burst test showing no fragmentation of tank

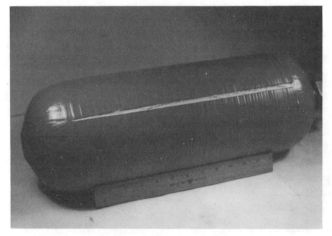

Fig. 13 - Exaggerated flaw growth test showing no increase in flaw size during pressure cycling

Fig. 14 - Impact test showing no
performance degradation

Fig. 15 - Internal examination of
cylinder with boroscope

Fig. 16 - Drag abuse test showing
no performance degradation

Fig. 17 - Metal liner assembly ready
for winding

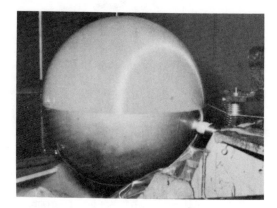

Fig. 18 - Initial filament windings
applied to liner

samples from each lot of 200
production units.

Some unusual tests have been
conducted to assure that units meet
specific customer requirements.
For example, a drag abuse test
(Fig. 16) was performed by attaching
a cylinder to a car, dragging it at
moderate speeds, hitting and
bouncing it on curb, and then
showing no degradation of performance by burst testing.

SPHERES - As pictured in
Figs. 17-22, spheres are manufactured
in much the same way as cylinders.
However, they are not now being
produced in large quantities and
are not DOT-authorized.

Fig. 19 - Partially completed
filament overwrapped liner

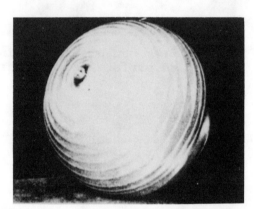

Fig. 20 - Pressure vessel after
winding and curing

Fig. 21 - Light weight composite
pressure vessel after
proof test

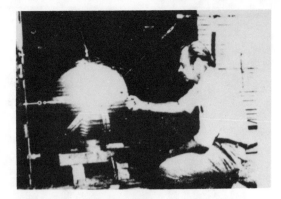

Fig. 22 - Performance test set-up

ADVANTAGES

The major advantage of composite pressure vessels is light weight; they weigh up to 50% less than comparable metal tanks. In addition, they are usually lower cost than high-performance all-metal aerospace tanks. Another important advantage is enhanced safety because of the leak-before-burst failure mode designed into composite tanks using fracture mechanics technology.

Other features of filament wound tanks include:

o Long life, low maintenance, rust free

o Proven in cyclic fatigue and static loading

o Compatible with cryogens, propellants, gases, liquids

o Non-shatterable under ballistic impact

o Demonstrated resistance to high humidity/temperature combinations, fungus, salt spray, sand, dust, sea water, sunlight

o Meet normal vibration, shock, acceleration and acoustic requirements

o Interchangeable with heat-treated steel, aluminum, nickel-base alloy and titanium vessels

USES

Composite tanks are ideal where light weight and high strength are important. Applications include containment of fluids for the following systems:

o Spacecraft

Propellant tank pressuri-
zation

Engine or instrument fuel

Engine purging

Auxilliary power

Astronaut life support

Cold gas thruster

o Missile

Guidance power source

Cryogenic instrument cooling

o Aircraft and Helicopter

Inflation of escape slide
and life raft

Emergency actuation of door,
landing gear and other
devices

High altitude gaseous O_2

Pneumatic engine starting

Emergency actuation of
escape seats

Hydraulic accumulator

Cryogenic instrument cooling

Engine compartment fire
extinguisher

Inflation of float for
landing in water

Pnuematic gun drive power

Crew walk around bottle for
in-flight breathing, and
individual passenger use

o Hydrofoil Ship

Fire extinguisher

Clearing of clogged water
inlet

Emergency actuation power

o Ground Service

Emergency breathing for fire
fighter, miner and rescue
worker

Automotive fuel such as
compressed natural gas or
hydrogen

Over-the-highway trailer
tubes for industrial
compressed gas

Medical or industrial compressed gas

Figs. 23-29 show typical applications.

Fig. 23 - Cylinder for fireman's
breathing apparatus

Fig. 24 - Cylinder for commercial
aircraft escape slide
inflation system

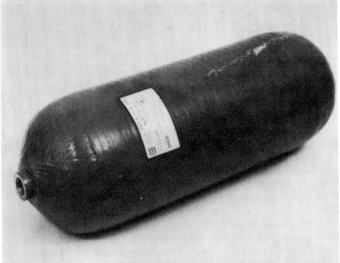

Fig. 25 - Cylinder for helicopter
flotation bag inflation
system

Fig. 26 - Cylinder for hydrofoil
missileship fire
extinguishing system

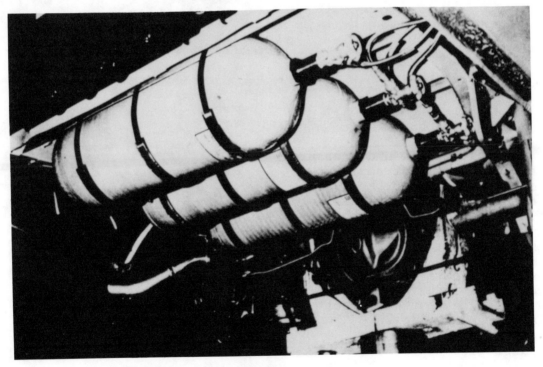

Fig. 27 - Experimental compressed
natural gas fuel tanks
for automobiles

Fig. 28 - Spherical gas storage tanks for aircraft auxilliary power system

SUMMARY AND CONCLUSIONS

Advanced filament wound pressure vessels are the result of applying aerospace materials and processes to manufacture useful commercial products. The transfer-of-technology has been successful by carefully retaining the essential parameters of the technology while modifying those that cause high costs characteristic of the aerospace industry. Thus, advanced technology but lower cost commercial composite pressure vessels are now available for military/aerospace applications.

Fig. 29 - Cylinder for manned maneuvering unit thruster system

CHAPTER 3

DESIGN AND ASSEMBLY

Reprinted from *Aviation Week & Space Technology*, McGraw-Hill, Inc., November 16, 1981

Space Shuttle Lightweight Tank Production Begins

By Edward H. Kolcum

Michoud, La.—Production has started on the first four lightweight external tanks for the space shuttle, and November is targeted as the deadline for changes in the initial block of advanced tanks.

The first large reduction in tank weight is being made as manufacturers and suppliers gear for volume production. At the same time, industry and government are searching for ways to reduce production costs.

Choke Point

Their objective is to eliminate the external tank as a potential choke point in space shuttle performance and launch scheduling.

The eighth production flight external tank will be the first lightweight model and is scheduled for delivery in Fiscal 1983. Five of the heavier tanks are planned for delivery in Fiscal 1982 to end production of this design. Lightweight tank delivery schedule calls for seven in Fiscal 1983, 11 in Fiscal 1984 and 19 in

Retrieval Device in Tests off Florida Coast

Diver-operated plug is lowered from the United Technologies Corp. recovery vessel Freedom in Port Canaveral, Fla., during a test exercise of the device that will be used for the final dewatering of solid rocket boosters after they are retrieved following space shuttle launches. The boosters, which are 149.2 ft. high, generate a maximum thrust of 5.9 million lb. They burn about 2 min., coast about 70 sec., and then are jettisoned from the space shuttle system. They are scheduled to impact 158 mi. downrange, where United Space Boosters retrieval specialists will insert ballast aerating retrieval booms into the nozzles. Compressed air is pumped through the booms and the motor casing tips from a vertical to horizontal position. The casings are then towed to Kennedy.

Fiscal 1985. At this time, volume production at a rate of 24 tanks per year is set to begin in the last quarter of Fiscal 1985. Planning is under way to increase this rate to 55 per year.

Actual production rates and deliveries will depend on the number of space shuttle flights that survive budget cuts. A 48-mission schedule through 1985 was reduced to 34 flights earlier this year (AW&ST May 11, p. 17). More recently, cancellation of 10 additional flights became a possibility because of a potential $500-million space shuttle overrun in Fiscal 1983 (AW&ST Aug. 17, p. 27).

Martin Marietta Aerospace, prime contractor for tank production at the Michoud Assembly Facility, has set a goal of reducing 6,400 lb. from external tank weight in producing the light model. The first flight tank weighed approximately 77,000 lb., about 1,000 lb. under National Aeronautics and Space Administration specifications.

The second flight tank weighed about 200 lb. less than the first.

Weight Reduction

To reduce tank weight by three more tons, the company made some changes based on structural test data, modified some fabrication techniques and materials and eliminated the exterior white paint (AW&ST Sept. 7, p. 19).

Structural changes included disposing of some stringers and tank ring frame stiffeners from the liquid hydrogen tank and modifying several support frames in this tank. Dome caps, which were chemically milled on only one side, will be milled on both sides. Changes were made in some materials to lighten solid rocket booster and aft orbiter attachment hardware.

The anti-geyser line has been removed and liquid hydrogen pressure line relocated to the opposite side of the tank alongside the liquid oxygen feed line. This allows ablator coating to be eliminated from a section of the tank.

The anti-geyser line was installed to provide a recirculation path and prevent formation of an oxygen bubble in the ullage area because of the temperature difference. It was found to be unnecessary. Cost/weight tradeoffs and studies indicate a number of other weight reductions that are possible but probably will not be done

Joint Soviet/French Space Mission to Be Launched Next Year

Work on Soviet/French space missions is under way in both the manned and unmanned regimes. Prime crew for a joint Soviet/French manned mission to a Salyut space station next year (lower left, facing page) are, from the left, Soviet cosmonauts Alexander Ivanchenkov and Yuri Malyshev and French cosmonaut Jean Loup Chretien. Backup crew for the mission (top, facing page) are shown undergoing contingency landing water survival training and are, from left, Soviet cosmonauts Leonid Kizim and Vladimir Solovyov and French cosmonaut Patrick Baudry. Launch is set for mid-1982. A mission already under way is the Arcad-3 spacecraft mission, also designated Oreol-3 (left). The vehicle was launched from Plesetsk on Sept. 27. A French-built antenna for wide-band telemetry is in the center of the end of the spacecraft.

for the first block of four lightweight tanks. The ground rule for the initial block was that if weight could be removed without compromising safety of flight and at a cost of $75/lb. or less, it would be considered. Additional candidates for weight reduction for later tanks include:

- Placing cables inside the external tank, thus removing the weight of exterior cable trays.
- Eliminating the liquid oxygen slosh baffle. There was less sloshing than anticipated on the first flight and data will be closely examined on slosh action during the remainder of the development flights. The baffle weighs 1,000 lb.
- Using composite materials.

Martin Marietta has come under pressure to reduce the cost of producing the external tank, the only expendable main component in the shuttle system.

A per-flight cost of $1.8 million was estimated for the external tank in 1971, and this cost held as recently as five years ago (AW&ST Nov. 8, 1976, p. 134). NASA today estimates that production tanks will cost $10.1 million each.

Robert W. Smith, manager of industrial engineering and production readiness for Martin Marietta here, said three changes in the external tank since the original specifications have caused the cost increase:

- Addition of the tumbling subsystem to insure the tank comes to Earth over an unpopulated area.
- Addition of the range safety system. Initially, it was believed that the external tank and solid rocket boosters could be destroyed with a single system located in the boosters.
- Broadening thermal protection system requirements.

The tumbling subsystem, which failed on the first launch Apr. 12, includes a pyro valve on the forward ring assembly of the liquid oxygen tank in a location near the top of the external tank.

Manifold Connection

The valve connects with a manifold that extends to a hollow lance inserted into the liquid oxygen tank. After separation from the orbiter, a detonator and pyro booster explode automatically, shearing the gate between the lance and valve. This allows residual gaseous oxygen to vent through the lance and valve and exhaust out a nozzle atop the valve, causing the tank structure to tumble end over end. It was believed debris in the valve on the first flight prevented it from operating. The system will be purged under a vacuum shortly before liftoff on subsequent missions.

The external tank range safety system is designed to destroy the tank on command using explosive charges. The system consists of two command antennas, two linear shaped charge assemblies, seven detonating fuses with two manifolds, a safe and arm device and two detonators. Ordnance is located in the liquid oxygen tank, liquid hydrogen tank cable trays and in the intertank structure. The antennas and batteries are located in the intertank. If a destruct command is received by either of the solid rocket boosters, it will activate the range safety system in the external tank and the other solid.

Thermal protection requirements represent the most radical change, and the present design results from a combination of experience, testing and a great deal of analysis, Smith said.

Some experience came from the first propellant loading test last January, when 88 sq. ft. of super light ablator material debonded when the adhesive shrunk because of cryogenic shocking (AW&ST Feb. 2, p. 18). Originally, the liquid oxygen tank walls were bare, exposing the 2219 aluminum alloy skin.

Now, the forward ogive area is covered with 0.38 in. of super light ablator, which in turn is covered with 2 in. of sprayed-on foam insulation. The latter material covers the remainder of the liquid oxygen tank, tapering from 2 in. to 1 in.

Icing Deterrence

Protuberances originally were not required to deter icing. Now, all protuberances and interface hardware are protected with thermal isolators, heaters or urethane foam covering. The original requirement called for 0.5 in. of foam insulation on liquid hydrogen tank walls. This has since been increased to 1 in.

Spray-on foam insulation is a polyisocyanurate fluorocarbon blown closed cell rigid material able to maintain the boiloff of −423F for liquid hydrogen and −297F for liquid oxygen. It also is stable to a 300F substrate temperature.

Super light ablator is a composite mixture of silicone resins filled with cork particles and silica and phenolic materials with heptane added to form a slurry.

Although 59,000 manhours in thermal protection system work were cut between the first and second external tanks, this procedure remains an extremely high labor intensive—and expensive—part of production.

The second external tank thermal protection work took 202,000 manhours, and the first consumed 261,000 manhours.

Thermal protection production improvements represent a key part of a Martin

Marietta commitment to reduce cost of the first 54 lightweight tanks by $66 million by investing no more than $20 million to pay for changes.

S. R. Locke, program manager-producibility for the external tank, said that, since the Martin Marietta cost reduction commitment was made two years ago, more than 175 viable ideas have been identified that would achieve about 70% of the savings. Of these, 36 are being implemented and will result in savings of $9.5 million, representing a 24:1 return on investment.

Another 44 candidates are being assessed and show a potential savings of roughly $20.6 million—a 6:1 return on investment. Some 65 additional ideas are being studied with potential savings of more than $38 million.

James B. Odom, manager of the external tank project at Marshall Space Flight Center, which is the NASA center responsible for the tank, said that while the saving may be $66 million on the first 54 lightweight tanks, it could total $800 million over the life of the program. For planning, a total of 491 tanks currently represents the program life.

Locke said the improvement program can involve changes in the design, process, method or materials of production, but it cannot change the form, function or fit of a component. "Beyond this, nothing is sacrosanct," he said.

One example is in molding super light ablator. Martin Marietta has developed an injection molding technique and the Massachusetts Institute of Technology has designed a screen molding process, both of which are replacing the hand molding technique now used.

The handpacking method used to fabricate sheets or shapes involves applying the ablator slurry by hand, inserting it in a vacuum bag where it is heat cured, and then machining it to dimension. Gas injection molding involves injecting the ablator slurry in a mold with gaseous hydrogen under a pressure of 150 psi. It then is heated and vacuum cured, reducing labor time 85% and saving 60% of the slurry which previously was wasted. The MIT process also involves use of a form on which the ablator slurry is pressed at 2-5 psi. and then heat treated.

"We are able to mold to net dimension, eliminating trimming by hand with plastic knives," Locke said.

Identified to be fabricated by the MIT method are 41 lightweight tank components; 114 components have been identified to be injected by the Martin Marietta injection technique for the first lightweight tank. Another 166 elements have been identified for the fourth and subsequent lightweight tanks. Potential savings for the net molding use are $10 million through the 54th lightweight tank and over $2 million in avoiding the cost of a new machining facility.

Another manufacturing improvement is a semi-automated weld inspection that will replace manual peaking and mismatch measurements made with a wire profile gauge, straight edge and optical comparator. A fully automated non-destructive examination device will determine not only peaking and mismatch but also buried and surface defects.

Potential savings of the semi-automated inspection system through the 54th lightweight tank total $244,000, and through the 491st external tank, $1.9 million, Locke said. The fully automated system could become available for the 12th lightweight tank or 20th flight tank. Through the 60th tank, potential savings are $5 million, and through the 491st tank, the potential savings are $97 million. Six individual teams are involved in assessing ideas for changes that will increase the producibility of the external tank.

Team Effort

Their disciplines are engineering, production operations, quality assurance, procurement, finance and planning. Candidate items have come from employees, contractors, subcontractors, suppliers and consultants. After an item has gone through the screening, approval, development and implementation phases, its actual use depends on concurrence by NASA and its ability to have positive effects on the program, long-term returns and short-term effects.

For example, a study is under way to replace the dual spray-on foam and super light ablator on the aft dome of the liquid hydrogen tank with a single spray insulation system.

The task will be undertaken in three steps:

■ Screening candidate materials by analyzing physical, mechanical and thermal performances, application and flammability.

■ Testing spray characterization and evaluating pilot manufacturing.

■ Final verification. With this new process, Martin Marietta

believes manufacturing flow time can be reduced by 33 hr. and shuttle payload capability increased by 260 lb. Six manufacturing steps will be eliminated with the single spray system, Locke said.

In the buildup to mass production, Smith said the facility is being designed with "assembly line geography so we can achieve assembly line mentality. The worker on the line needs to know the line is moving."

Processing is heavily computerized and a production simulation model is determining such factors as work station capacity and if different tools can increase that capacity, what tools are likely to fail, logistics, schedules and workarounds.

Test Articles

Odom said tools and plant layout were developed with the idea in mind that they would be used to build test articles as well as external tanks at the maximum rate. Experience has shown that some components should be worked offline with tools that cost a fraction of those on line, he said.

A computer program has been developed for performance control and measurement that comprises work standards and simplified work papers, tool facility status, procurement and vendor data, and an integrated parts status system. This system maintains an inventory status, locates each tank part and status, and logs manhours spent on each manufacturing function.

Smith said that at the same time manufacturing is being expedited, product quality must be maintained. "We must keep

track of the origin of each batch of materials.

Through trend analysis, we can find out where we can minimize inspection and maximize quality." He said Martin Marietta is automating this information as well as supplier nonconformance and other factors in quality control.

Manufacturing Goals

In manufacturing, Smith said three goals have been developed:

■ Reduce support labor. This is being done by establishing standards, manloading the detailed workplan and measuring performance.

■ Decrease time delays for touch labor. Smith said this requires an assembly line approach and discipline, improving floor planning and control, providing on-floor support by the Defense Contract Administration Services, NASA's quality agent in Michoud, providing on-floor support by the material review board, a joint Martin Marietta-government group enpowered to authorize changes, appointing a policy board on-call within 2 hr., providing authority to proceed with touch labor or to schedule delays, situating managers on the assembly line floor and having a consolidated inventory.

■ Increase production speed. The emphasis will be on design-to-build sequencing, improving production paper, establishing detailed standards and labor reporting, issuing authority for unplanned standards and improving manufacturing labor performance reporting with updated plans and action performance.

Martin Marietta has approximately 2,000 subcontractors and suppliers in the project, including:

■ Aluminum Co. of America—Forgings including solid rocket booster attachment fittings, ball forgings, longerons, forward ogive forgings and diagonal struts.

■ Reynolds Metals—Machined aluminum for liquid hydrogen tank barrel panels.

■ Kaiser Industries Corp.—Machined aluminum for liquid hydrogen tank barrel panels.

■ Kaman Aerospace—Slosh baffle segments.

■ Aerochem—Liquid oxygen tank barrel panels.

■ Aircraft Hydroforming, Inc.—Hydraulically pressed gore and ogive panels. Outer, inner and intermediate chords. □

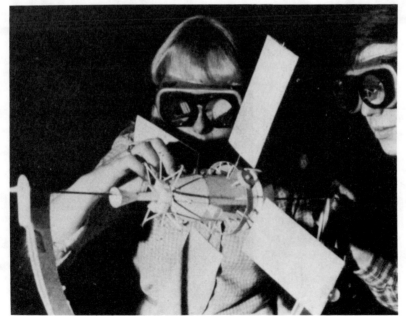

Satellite to Attempt Theory of Relativity Verification

Marshall Space Flight Center engineers conduct a solar array sun shadow analysis on a model of the Gravity Probe-B satellite, which will attempt to verify the theory of relativity. The analysis helps determine the size and shape of solar arrays that will be designed to provide a continuous one-year power level to the spacecraft.

Presented at the SME Automated Fastening Conference, October 1984

Automated Fastening Large Structures
A New Approach

by Donald F. Lumley
Martin Marietta Aerospace

The Space Shuttle is America's economical and effective Space Transportation System (STS) developed by the National Aeronautics and Space Administration (NASA) to conduct space missions for projected national and international space program activities. The Space Shuttle vehicle consists of three major elements: a reusable manned Orbiter, an expendable External Tank (ET) containing the Orbiter propellant, and two reusable Solid Rocket Boosters (SRBs). The External Tank, built by Martin Marietta Aerospace at the NASA Michoud Assembly Facility in New Orleans, Figure 1, is the largest element of the Space Shuttle. It serves as the backbone structure for attachment of the Orbiter and SRBs and, also, contains and delivers propellants for the three Orbiter main engines. The External Tank accommodates the complex stresses created by its own weight and that of the Orbiter prior to launch, then the thrust generated by the Orbiter and the SRBs during launch. The overall Shuttle has a gross lift-off weight of 4.5 million pounds. Fuel can be supplied from the External Tank to the three main engines at the rate of 45,283 gallons of liquid hydrogen (LH2) per minute and 16,800 gallons of liquid oxygen (LO2) per minute.

Figure 1. Completed External Tank Being Transported From Michoud Assembly Facility

The External Tank is 153.8 feet long and 27.6 feet in diameter. It weighs approximately 69,000 pounds empty and when loaded with propellants at launch weighs approximately 1,660,000 pounds. Three primary structures make up the ET; a LO2 Tank, an Intertank, and a LH2 Tank. Both propellant tanks are constructed of aluminum alloy skins with support or stability frames as required, and their skins are butt fusion welded to provide reliable sealed joints. The Intertank aluminum structure utilizes mechanically fastened skins and stringers with stabilizing frames. The External Tank primary structure is shown in Figure 2.

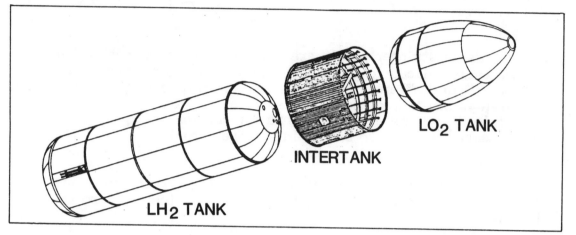

LO$_2$ TANK

INTERTANK

LH$_2$ TANK

Figure 2.   External Tank Primary Structure

The Intertank, Figure 3, is the ET structural connection that joins with both the LO2 and LH2 tanks to provide structural continuity between these assemblies. Its primary functions are to receive and distribute all thrust loads from the SRBs and transfer loads between propellant tanks. The Intertank also functions as a protective compartment for housing instrumentation, range safety components, and other subsystems.

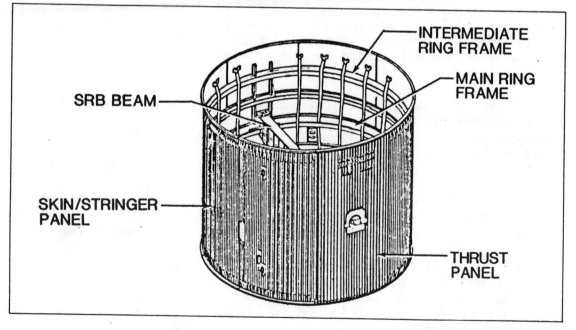

INTERMEDIATE RING FRAME

MAIN RING FRAME

SRB BEAM

SKIN/STRINGER PANEL

THRUST PANEL

Figure 3.   Intertank Structure

The Intertank cylindrical structure consists of two integrally machined thrust panels and six mechanically fastened stringer stiffened panels. It is 27.6 feet in diameter and 22.5 feet long. The two thrust panels distribute the concentrated axial SRB thrust loads to the LO2 and LH2 tanks and adjacent skin panels. The thrust panels are selectively machined with tapered skin thicknesses and external integral ribs. The six stringer stiffened panels are similar to each other except for penetrations, system installation provisions, and an access door. Each panel is 10.8 feet wide and 22.5 feet long and includes a forward and aft chord for attachment to the LO2 and LH2 tanks.

The skin/stringer panels are each made of two aluminum skins mechanically spliced longitudinally by internal and external butt straps. Skin doublers provide necessary reinforcement for areas where the skin is penetrated and localized reinforcements to distribute thrust loads. There are 18 aluminum hat section stringers mechanically fastened to each skin/stringer panel.

The six stringer stiffened panels and two thrust panels are mechanically spliced using longitudinal butt splices to form the Intertank skin shell.

One main frame is employed to transmit the transverse SRB thrust loads and the intermediate ring frames stabilize the cylindrical shell. The main frame is constructed of machined outer and inner chords joined to webs to form an I-beam 20 inches deep. The four intermediate ring frames are constructed similar to main frame and are 12 inches deep.

The SRB beam assembly running through the middle of the Intertank is a rectangular box beam. It is 42.95 inches deep at the center, tapers to 26 inches at the ends, and is 15 inches wide. Two SRB thrust fittings, machined aluminum forgings, are attached at either end of the beam and provide for SRB attachment.

The Intertank structural assembly is performed at the NASA Michoud Assembly Facility (MAF) in New Orleans, Louisiana. Major components including thrust panels, stringer panels, frame quadrant sections, and the SRB beam are subcontracted.

The Development and early production Intertanks were built on modified Saturn S-1C tooling remaining from the previous Saturn/Apollo program at Michoud. A new Intertank manufacturing approach was developed utilizing a half section subassembly. New tooling was designed and built and implemented in a manual mode in the fourth quarter of 1983. The heart of the half section manufacturing approach will be a new and unique Automatic Riveting System for large structures. The new riveting system is being built by GEMCOR and is currently in preliminary checkout at their facility. It will go into operation at Michoud in the third quarter 1985.

The Intertank half section, Figure 4, consists of three skin/stringer panels, 180 degree sections of the five frames, butt splices, and miscellaneous items. The components are joined using blind fasteners (under hat section stringers), 1/4 inch diameter A286 Hi-Sets, and Hi-Loc fasteners.

Figure 4.   Intertank Half Section On Transportation Dolly

The first position in the half section family of tooling is the panel and frame tack fixture. In this fixture, the four 180 degree intermediate frame segments, 180 degree main frame segment, the three 45 degree skin/stringer panels are positioned and located for tack fastening. Blind fasteners located under the hat section stringers are the primary tack fasteners.

The third position is for finish, inspection, repair and bracket and subsystem support structure installation. In the current manual mode, fasteners that will be installed by the Automatic Riveting System, are installed in this fixture or the tacking fixture.

These two tooling positions as well as the foundation foot-print for the Automatic Riveting System are shown in the photograph, Figure 5.

Figure 5.   Footprint For Intertank Automatic Riveting System Shown Between First and Third Tooling Positions

The second position in the family will be the Automatic Riveting System, Figure 6, which will install the fasteners to secure the skin panels to the frames and complete the butt splices. The majority of the fasteners used in this operation are 1/4 inch A286 Hi-Sets.

SYSTEM DESCRIPTION

The Automatic Riveting System utilizes a new and unique system of driving the drill/rivet heads under computer numerical control on vertical inner and outer columns with the work piece rigidly supported on a 360 degree rotary positioner. This Vertical Drivmatic is capable of installing any one piece fastener, including slugs, up to 3/8 inch diameter without displacing the workpiece. The inner to outer head load bias to the workpiece is limited to a maximum of 50 pounds under the heaviest riveting conditions.

The system is capable of riveting structures with both internal and external stiffening. The rivet installation rate exceeds twelve fasteners per minute in the dual head mode and seven per minute in the single head mode. The key features of the system are described in the following paragraphs:

VERTICAL DRILL/RIVET COLUMNS - ROTARY POSITIONER

The columns are 33 feet high and measure eight feet by ten feet at the base. Two sets of opposing carriages, Z axis, are positioned by electro servomotor rack and pinion drives with dual hydraulic motor counterbalances. This arrangement permits 28.5 feet of vertical travel or 20.7 feet of working range with either set of carriages parked. Each pair of carriages are controlled by separate CNC systems The inner column carriages (slaves) provide synchronous movement to the outer carriages and are positioned accordingly..

The outer Z-axis carriage houses the Y-axis carriage which, in turn, houses the transfer head. The transfer head carries the drill spindle, hole inspection system, and bucking ram and is actuated by the transfer cylinder. The outer Z-axis carriage also mounts the automatic tool changer, automatic injector changer, vision system and TV camera. Six inches of in and out head travel, Y axis, is provided for clearing external stiffeners and accommodating work plane variations.

The inner column carriage contains the upset ram which has 24 inches of programmable retraction and programmable rotation of 350 degrees about the rivet centerline.

Figure 6.   GEMCOR Automatic Riveting System During Preliminary Checkout.

Plus or minus one-half inch X - axis translation is provided in each carriage to permit independent edge margin control.

The CNC rotary positioner, C axis, is a 28 foot diameter ring shaped rotary table which positions the part holding fixture and work piece for the drill rivet operations.

CONTROL SYSTEM

The Control System includes the operator's console; two Allen-Bradley 7320 Computerized Numerical Controls (CNC); three Allen-Bradley Programmable Logic Controllers (PLC 2/30); and the General Electric Vision System. The controls are interfaced to permit independent and simultaneous operation of both sets of heads. The control system block diagram is shown in Figure 7.

Primary control of the Vertical Drivmatic is accomplished through utilization of two CNC controllers. One controls the Upper Head, while the other controls Lower Head movement. With the dual CNC system one will be considered the master, which controls the 'C' Axis

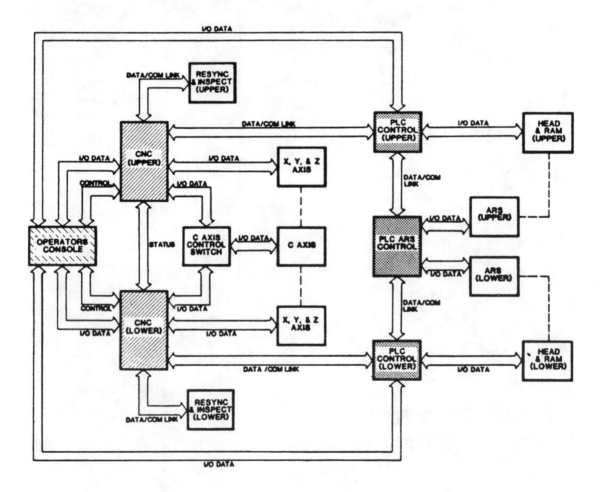

Figure 7.   Control Systems Block Diagram

programming and is responsible for locking out any head motion that could cause workpiece damage. Either CNC can have the capability of controlling the C Axis, but only one at a time will do so.

Head functions will be controlled by two PLC's in communication with the CNC. In addition a third PLC will be used to control both ARS Systems. The Tack Resynchronization Visual System provided for each head will interact with its associated CNC.

The Operator's console houses all operator controls in a single main control panel. The two Allen-Bradley CNC main control panels with the CRT display and alpha-numeric keyboard are located in the operator's console along with all controls and indicators necessary for the operation of the system and riveting process. The Operator's Console, shown in Figure 8 also houses the color TV monitors and vision system displays for each set of heads.

The system is capable of controlling the riveting process by either servo upset force or head height control. The mode of operation as well as the upset force or head height can be selected from the console or the parts program.

Figure 8.   Operator's Console Provides Independent Control of Dual
Drill/Riveting Systems

RIGID WORKPIECE LOCATION AND CLAMP

The Automatic Riveting System locates the work plane of the rigidly mounted Intertank half section and completes the riveting process without displacing the work piece. The following explains the GEMCOR developed method of applying rivet upset forces while putting less than fifty pounds of bias force on the workpiece. The Y-Axis carriage moves in under system pressure. When the pressure foot, which is at low pressure, touches the work surface and yields the pressure foot encoder senses movement and stops the Y-Axis carriage. The Y-Axis then retracts until the encoder indicates home position. The pressure foot is now fully extended and can be pressurized with system pressure.

The inner ram now advances until the clamp makes contact with the work surface. The clamp then switches from low to high pressure and the work piece is fully clamped. The Drivmatic Drill and riveting process is then initiated.

EDGE MARGIN AND TOOLING HOLE RESYNCHRONIZATION VISION SYSTEM

The vision system has the capability to detect the edge of a part and shift rivet location for a single hole or group of holes using X-axis translation so that a predetermined minimum margin is maintained. This mode of operation will be used to locate the edge of hat sections when riveting Intertank panels to frames. In the hole resynchronization mode, the vision system has the capability of resyncing the parts program to a tooling hole. The resynchronization is accomplished by means of X-axis translation and/or Z-axis move. The system also has the capability of resyncing on a second tooling hole, computing the offsets for intermediate fasteners, and shifting the fastener locations accordingly.

AUTOMATIC TOOL CHANGER

An Automatic Tool Changer on each outer carriage changes the drill and bucking anvil by means of the CNC parts program or operator console input. The tool changer has the capability of accomplishing the tool change without moving the workpiece. The GEMCOR developed Automatic Tool Changer has 18 tool positions comprising six bucking anvil positions and 12 drill positions. An Automatic Injector Changer accommodates up to five injectors.

The Automatic Tool Changer can complete a drill change in less than 30 seconds and a bucking anvil and injector change in 30 seconds. The Automatic Tool Changer is shown in Figure 9.

Figure 9.   Automatic Tool Changer Holds 12 Drills and 6
Bucking Anvils

AUTOMATIC RIVET SELECTION SYSTEM

An independent Automatic Rivet Selection System (ARS) is provided
for each set of riveting heads. Each ARS has a floor mounted feed
station containing 16 vibratory bowl feeders for unscrambling and
aligning the rivets. The rivet selection is controlled by the CNC
using a miscellaneous (M) function, an automatic stock thickness
measurement system, or through the operator console. Selected rivets
are automatically dispensed one at a time, blown through the injector
tube to the injector and fed into the drilled holes as part of the
automatic riveting cycle. A backup or reserve feed capability is
included to permit manual feed from three individual drop tubes for
each riveting head. The ARS, which can be expanded to include
additional feeders, is shown in Figure 10.

Figure 10.
Automatic Riveting Selection
System Selects & Feeds
Fasteners From 16 Vibratory
Bowl Fasteners

AUTOMATIC INSPECTION

A programmable probe type hole inspection system is incorporated in the outer head which performs a plug gage type inspection. A precision conical probe is inserted into the drilled hole and its displacement measured with a Linear Variable Displacement Transducer (LVDT). From the linear displacement data, the hole diameter is computed and compared to the established tolerances prior to rivet insertion. The system is interfaced with the CNC system to provide the capability of interrupting the drill/rivet cycle and displaying an operator message on the CRT when a drill change becomes necessary to maintain hole tolerance.

A rivet process data system provides real time hard copy printout of sequence number, upset force, hole size, and rivet head height. The frequency of printout is programmable to provide data as desired from every rivet cycle down to an exception only (out of tolerance) basis.

The system is capable of programming the hole size, upset force, and head height tolerances and the frequency of hole size inspection. The basic programming of the frequency of hole inspection is not accessible from the operator's console, however, the operator has the capability to selectively add inspection of any holes without changing the basic program.

SUMMARY

This Automatic Riveting System, being procured from GEMCOR by Martin Marietta Aerospace for NASA will incorporate a new and unique automatic riveting system approach. It will bring to the Michoud Assembly Facility and External Tank Program the latest automatic riveting technology permitting further ET cost reductions and capability for low cost production of large structures for future space programs.

Reprinted from *Manufacturing Engineering*, March 1980

Skin Milling for the Space Shuttle

The Space Shuttle poses an unprecedented array of engineering problems. High in priority among them is the problem of developing a skin — and skin material — capable of withstanding the heat of reentry. The author, a vice president of SME, tells how it's done

ROBERT L. VAUGHN, CMfgE, PE
Lockheed Missiles and Space Co.

THE NEXT BIG STEP in the investigation of outer space will be accomplished with the Space Shuttle. Upon completion of its final tests, this craft will be able to take off vertically, but otherwise perform much like a conventional airplane. After take-off, it will reach a velocity and altitude that will enable it to orbit the earth much like the space capsules of the Gemini and Apollo programs.

There will be one significant difference, however. The Shuttle will be able to return to earth and then repeat the process at a later date. This flight to space and back will be repeated numerous times with the same shuttle, and therein lies a problem — that of preventing the burn-up normally associated with capsule reentry.

Advent of LI 900. Lockheed Missiles and Space Co. has developed a protective material that can do the job. It's an all-silica material called LI 900 (Lockheed Insulation — 9 pounds to the cubic foot). In the space shuttle application, LI 900 is machined into tiles of varying shapes, sizes and thicknesses. These tiles are used to protect more than 70% of the shuttle's exterior surface. The remaining surface, which is exposed to temperatures as high as 2865°F (1574°C), is protected by carbon-carbon and various other materials.

One of the more unusual properties of LI 900 is its ability to dissipate surface heat rapidly while retaining interior heat through slow heat transfer. As a result, it is possible to handle a piece of this material by holding it by the edges seconds after it has been removed for a kiln heated to 2500°F (1371°C).

Importantly, the LI 900 material is designed to withstand many exposures to the high temperatures encountered during reentry. As such, it is expected to make the shuttle reusable for as many as a hundred flights.

Made of Sand. LI 900 is produced from short-staple, 99.7% pure amorphous silica fiber derived from common sand. Through a series of complex processes, the sand is converted into a cotton-like fibrous material. Subsequently, it is put through another series of processes until it emerges in its final form as thermal insulation.

The fibers extracted from common sand are purified, cleansed, dried and weighed into controlled batches. These batches are put through a continuous casting process and made into LI 900 blocks called "production units."

Casting the Blocks. The casting process begins with a batch of fibers which are suspended in deionized water. The mixture is then cast into a plastic mold and water is partially extracted. Precise amounts of fiber binder solutions are introduced into the mold before the block is ejected from the mold onto a conveyor.

The wet blocks, weighing about 29 lb (13 kg), are conveyed to a tunnel microwave oven which removes their excess moisture. They are then transferred to a tunnel kiln and sintered at 2350°F (1288°C).

Next, the blocks are saw-trimmed to remove skin crusts. At this point, they are in the form of production units with volumes of 0.5 ft³ (0.01 m³) and weights of approximately 4 lb (1.8 kg). To verify homogeneous density, sample blocks are subjected to X-ray examination. They are then cut into flat blanks which are finally machined into various sizes and shapes of insulation tiles. The tiles are then coated, final machined and put into arrays for installation on the space shuttle by Rockwell International.

Computerized Drawings. Since each of the 32,000 tiles used in a space shuttle differs dimensionally from all others, the usual method of making a drawing and controlling changes for each tile would be extremely expensive. Rockwell International, the prime contractor, has devised an engineering system which uses computerized data in combination with a limited number of engineering drawings to define the dimensions of each tile and its precise location and tolerances in the shuttle. Rockwell International furnishes the master dimensions in the form of magnetic computer tapes and a master dimension data book which defines each tile's geometry. This is supplemented with engineering drawings that define the location and tolerance for various major structural areas of the shuttle.

Lockheed translates these tapes and

A WHITE HOT BLOCK OF LI 900 is held manually by a technician in this photograph. The secret of his ability to hold this material lies in LI 900's slow rate of heat transfer from its interior, but extremely fast rate of heat dissipation at its edges.

drawings into computer programs for machining and measuring each tile and for the respective groups of tiles in an array for installation. The master dimension data on magnetic tapes give a series of points along station lines of the shuttle, approximately an inch or less apart, depending on the contour in question.

The master dimension manual covers major sections of the shuttle's structure such as wing, fuselage, and vertical stabilizer. Engineering drawings provide the data required for installation of array frame assemblies, i.e., tile groups in frames. Tile and array boundaries are identified by either station planes or master dimension identifiers in the master dimension manuals. The drawings also identify gaps and tile-to-tile step tolerances.

drawings. Lockheed has developed an array concept in which each array contains an average of about 22 tiles. The tiles are locked together in their precise positions for final machining and attachment. There are about 873 array frames for the shuttle. Each can be reused a number of times.

An array frame assembly is comprised of a basic plate vacuum chamber contoured to duplicate the outer mold line (OML). The OML describes the boundaries of the specific groups of tiles. The basic function of the array frame assembly is (1) to orient the group of tiles and hold them securely under vacuum during machining of the inner mold line and (2) for use in subsequent installation on the vehicle. The installation is accomplished by adhesively bonding the tile to a strain isola-

tor pad which is then bonded to the shuttle's aluminum skin. RTV-560 adhesive is used for both bonds. Finally, the array frame serves as a shipping container. In all, the job of fitting tiles together on the shuttle's skin is somewhat like assembling the world's largest three-dimensional jigsaw puzzle on a surface twice the size of a basketball court.

The outer mold line and inner mold line of the array, as well as the boundaries of each tile, are programmed for NC machining via Rockwell's master dimension computer data, which has been converted for use on Lockheed Computers. The sizes of the flat blanks from which the tiles are made are also established by these sets of data. NC drafting machine plots are also generated by Lockheed's computer group for each array, depicting the geometry in three-or-five-axis modes. These plots are the universal references used by designers, planners, NC programmers, and by inspectors in the final array assembly area.

Computer Augmented Manufacturing. Lockheed has also devised a unique plan for machining both tiles and array assemblies, using a completely automated computer augmented manufacturing (CAM) process. The system consists of five specially designed NC machines equipped with computer numerical control (CNC). Three of the machines are five-axis units, two are three-axis units. The CNC is tied to an on-line direct numerical control (DNC) system which, in turn, is supported by six disc drives. These drives are capable of retaining all the machine control data necessary to produce a complete shipset of tiles.

The machine control data stored on

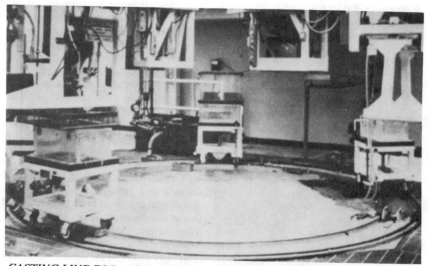

CASTING LINE FOR LI 900. Movement is counterclockwise. Molds are loaded at the left. Agitated fibers are loaded in the mold at the far side of the line. Completed cast blocks are automatically ejected at the far right.

Basic Data. Because of the complexity of their geometry, some tiles are not made by the master dimension systems. Supplemental dimensional data are furnished by individual tile drawings.

In general, there is not a single document that gives the complete dimensions of the finished item. These data, which define the geometry of each tile required for installation, must be adjusted for shrinkage allowances for machining and coating. Shrinkage characteristics of LI 900 are not linear and depend on the geometry, thickness and fiber lot of each tile. Lockheed Engineering reviews the tile configurations and issues a series of drawings which direct the NC programmer to incorporate the correct shrinkage allowances.

The same basic data also goes to the array frame assembly design group. This group reviews the Rockwell data and generates array frame assembly

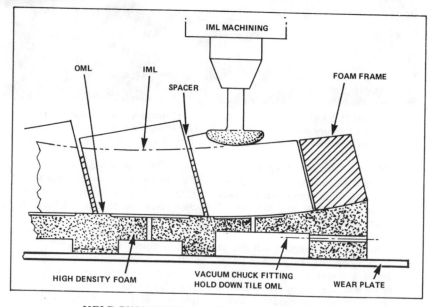

HELD BY VACUUM, the tiles are machined in an array fixture with specially made diamond tools. Contour being machined is the inner mold line.

these disc drives are identified by individual part number. The data are called by the NC machine operator, using a precoded punched IBM card. This card contains the part number, coating class of the tile, blank size, machine type, mirror program, and so on. The part number accesses the data from the DNC system and loads the data onto a floppy disc at the CNC control.

The fiber lot identity triggers the minicomputer at the CNC unit to select the proper cutting compensation for shrinkage allowances. The mirror image data informs the system to produce either a right-hand or left-hand part. All of these features, which are unique, were specially designed for Lockheed by the Allen-Bradley Co. and the Onsrud Division of the Danly Machine Corp.

Special Cutters. Confronted with the need to produce thousands of different tiles and array assemblies, Lockheed found it necessary to develop special cutters dimensionally capable of holding the exacting tolerances and surface textures required.

LI 900 is easily machined with any work on LI 900. Tool criteria were established as follows:

▶ Diamond size 60/80 filled with 80/100 mesh.

▶ The diamond-coated surface must be free of irregularities or voids. No diamond or plating may project above the nominal surface of the tool.

▶ Diamond must cover approximately 75% of the cutting surface.

▶ Individual diamonds must be well retained, but must not be completely covered with plating material.

▶ The tool-diameter tolerance is ±0.002″ (0.05 mm).

▶ Runout between the diamond-coated diameter of tool and shank must be held within 0.002″ (0.05 mm).

▶ The minimum corner radius is to be 0.04″ (1.0 mm).

Speeds and Feeds. Testing to determine optimum spindle speeds and feeds with these cutters showed that both could be as high as the machine tool was capable of producing. A one-inch (25.4-mm) cutter was run at 14,000 rpm, full width, at five inches (127 mm) depth at a rate of advance of 120 inches (3048 mm) per minute.

2300°F (1260°C) on reentry — are black and called ''High-temperature Reusable Surface Insulation.'' Low-temperature tile — coated with white and exposed to reentry temperatures of 1200°F (649°C) — are called ''Low-temperature Reusable Surface Insulation.''

Coating material is essentially a highly purified frit with binders. The slurry for the black coating utilizes denatured alcohol while the white coating is suspended in deionized water.

Final Inspection. Since the final geometry of each tile is not defined by a single document, as is the common practice with most machined parts, it was necessary for Lockheed to develop an inspection method. The method uses the computer master dimension data and drawings provided by Rockwell International.

Lockheed Engineering reviews and generates a mathematical model of each tile for inspection by the Bendix Cordax 5000 and 1000 measuring system. The Cordax uses a Digital Equipment Corporation PDP-11 computer. The procedure used is as follows:

Twenty-five measuring points are established — five on each of the four sides and five on the outer mold line surface — for each tile. These data are then forwarded to Quality Engineering where they are loaded into the memory banks of the computer system tied to the Cordax. When a tile is set up on the Cordax and probed, the actual measurements are recorded and compared to the standards in the memory banks. The on-line system immediately accepts or rejects the part and prints a hard copy showing the deviations found.

Unlike the individual tiles, the array frames are inspected directly on the NC machine with a pneumatic probe system. As with the tiles, deviations are recorded and retained by Quality Engineering. The inspection system is keyed to Rockwell's master dimension data, which also develops the NC machining programs after being processed into the Lockheed system.

Progress to Date. The basic design of the advanced computer-aided manufacturing methods are now well established and are being used on a production basis in the manufacture of high-temperature reusable surface insulation tile. Almost 120,000 tiles will be manufactured for the four-shuttle program.

Any assessment of the overall program must conclude that the methodologies used are making it possible to complete this next phase of man's continuing exploration of space. Equally important, the technologies themselves represent a significant step forward in the continuing evolution of American industry. ■

ARRAY FRAME VARIATION is shown in this photograph. Outer and inner mold lines in the array are machined by NC. Data for programming the NC machines are obtained from Rockwell International's master dimension computer data.

type of cutter under cutting loads of two to eight pounds (8.9 to 35.6 N). However, its silica composition makes it extremely abrasive, and high-speed or tungsten carbide cutters are quickly dulled.

The machining operation selected is similar to a grinding operation in that it employs an abrasive coated cutter. These cutters were developed by Lockheed — a necessity, since commercially available diamond-coated wheels were found to be unsatisfactory for close tolerance

An 8000-rpm spindle speed was selected to optimize cutter wear and tile finish. Currently, about 43 different cutter sizes and shapes are being used to generate the tile configurations.

Finishing the Tiles. When all machining operations on a tile's outer mold line have been completed, the tile is coated, sintered and waterproofed. Two classes of ceramic coatings are used — one for high-temperature tile, the other for low. The high-temperature tiles — exposed to temperatures of

Reprinted from *Welding Journal,* July 1971

Shuttle orbiter concept

Brazing of Refractory, Superalloy and Composite Materials for **Space Shuttle** Applications

involves the evaluation of several different alloys to fabricate orbiter heat shields and an aluminum-boron tape intended for use in metal-matrix-contoured composite structures

BY C. S. BEUYUKIAN

ABSTRACT. The shuttle orbiter heat shield will be fabricated from various materials, both metallic and nonmetallic. Non-metallic areas are expected to reach temperatures as high as 3200° F, while the metallic portions will be subjected to temperatures approaching 2500° F.

Work reported in this paper pertaining to the orbiter heat shield is confined to the metallic portion. Five different metals are being evaluated. The materials are TD-Ni-Cr, Cb-C129Y, Cb752, Haynes 188, and Inconel 625.

Joining these materials into panels for heat shield applica-

C. S. BEUYUKIAN is a Project Engineer, Advanced Manufacturing Technology and Development, North American Rockwell Corp., Downey, Calif.

Paper presented at the Second International AWS-WRC Brazing Conference held in San Francisco, Calif., during April 27-29, 1971.

tions is of major concern. Brazing is an established and reliable method of joining thin-membered assemblies and is characterized by certain advantages in fabricability. Use of the vacuum-furnace brazing process is being investigated as a primary or supplemental process to join materials for use on the shuttle orbiter heat shield.

Brazed aluminum-boron metal-matrix-contoured composite structures are also being evaluated for shuttle orbiter applications. These high strength structures are viewed with considerable interest and may be used extensively.

Introduction

Up to now, men and systems have been transported into space by expendable booster systems, systems used for a single mission. In order to reap the full benefits of space, we must conceive and put into use a reusable transportation system, one that will dramatically lower the overall cost of present-day launch systems. This should be realized, so far as possible, within the present technology and with the use of existing facilities. Its maintainability should be such that it can be used for 100 missions.

The shuttle was conceived as such a system. Major phases of its mission are boost flight, orbital operations, orbiter and booster entry, and ground operations. Basically, the mission requires that the shuttle craft be launched from a vertical position, with the orbiter riding piggyback atop the booster, to an altitude of about 200,000 ft. There, about 3 minutes after liftoff, the orbital vehicle separates, and the booster returns to the launch area.

Current studies call for the orbiter to be injected into a parking orbit of about 50 to 100 nautical miles, then into a 250-nautical mile orbit. Upon completion of its space tasks, the orbiter departs its orbital sphere, crosses through the supersonic high-temperature region, slows to subsonic speed, and levels off to land like a jet aircraft.

The shuttle is the immediate challenge of space. To help answer this challenge, the National Aeronautics and Space Administration, Manned Spacecraft Center, has awarded design and planning study contracts for a shuttle vehicle system. Flight status could be reached late in the decade. North American Rockwell is one of the aerospace firms awarded such a Phase B contract. The work described in this paper, funded by the company as shuttle-related research, applies specifically to the orbiter.

Heat Shield Materials

The shuttle orbiter's heat shield will consist of various materials, both metallic and nonmetallic. Some portions of the shield are expected to reach a temperature as high as 3200° F. The current baseline concept for most heat shield areas requires the use of reusable exterior insulation (REI). The use of metallic high-temperature materials for heat shield application is currently intended as a backup system for the REI. Studies indicate that the metallic portions of the heat shield will be exposed to temperatures approaching 2500° F.

What materials will be used for these metallic portions and what are they expected to do? Several metals are considered as candidate materials for the high-temperature areas. Among them are TD-Ni-Cr, Cb-C129Y, Cb 752, Haynes 188, and Inconel 625. Generally, TD-Ni-Cr can be defined as a nickel-chromium-based superalloy strengthened by an ultra-fine and highly uniform dispersion of thoria in the matrix. This material is being considered for use at temperatures up to 2000° F. Columbium C-129Y is a high-strength refractory columbium alloy developed primarily for reentry vehicles. It is currently being produced by the electron-beam fabrication process. This material is reported to have excellent thermal stability and coatability. Columbium 752 was developed principally as a high-strength alloy with good fabricability. The 10%

Fig. 1—Formed corrugation of Haynes 188

tungsten content reportedly enhances the elevated-temperature properties of this material. Both of these columbium alloys are being considered for use at temperatures from 2000 to 2500° F.

Haynes 188 is one of the newer cobalt-base, high-temperature, high-strength alloys. It is reported to have outstanding oxidation resistance. This alloy is being considered for use in heat shield applications at temperatures up to 1800° F. Inconel 625 is a nickel-chromium alloy reportedly having excellent resistance to corrosion and oxidation as well as good strength and toughness at high temperatures. This alloy is being considered for use at temperatures up to 2000° F. All these metals are being considered for use as heat shield materials in specific areas of shuttle orbiter vehicles, as determined by the temperatures in the areas.

Joining Methods

How will these materials be joined? No single joining process can be expected to be used in the fabrication of the many engineering materials being considered. Each joining process exhibits advantages or disadvantages that either enhance or restrict its use. Configuration, cost, end use, and actual service environments all influence the choice of joining processes for specific applications.

Currently, joining methods for certain of these materials are rather restricted. This may be because of one or a combination of factors. Prime factors are:

1. Adverse effects upon the base metals.
2. Limitations to service environments.

Fig. 2—Details for Haynes 188 pi-strap assembly

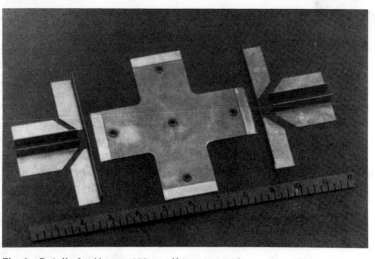

Fig. 3—Details for Haynes 188 cruciform assembly

3. Lack of knowledge with respect to engineering design requirements and manufacturing technology and capabilities.

Design considerations for joining these metallic heat shield materials are aimed primarily at two of the common processes: resistance spot welding and mechanical fastening. Unfortunately, each process has shortcomings when the full spectrum of usability is considered.

Joining of these superalloy and refractory materials is of major interest, even concern. A review of all joining processes reveals that the brazing process—especially high-temperature brazing—has definite economic advantages over many other methods. Fabrication of thin-membered assemblies, such as required for heat shield applications, makes brazing extremely attractive and promising.

Metal Matrix Composites

Current thinking for space shuttle reentry concepts includes the possible use of the new metal matrix composites. They are looked upon as candidates for lightweight, high-strength, structural stiffeners and auxiliary components. A most recent development is a concept that uses aluminum-boron tape. It is also being seriously reviewed for application to the shuttle orbiter. This 26 in. long composite structure, in the shape of a hat section, consists of 50 separate pieces of tape details and has been vacuum-retort-brazed in an isostatic gas-pressure furnace. The filaments are oriented in four directions.

Scope of Paper

This paper reports some of the work performed, or in progress, on the metallic heat shield materials. The specific objective of the program was to investigate and develop the vacuum-furnace brazing process, and its techniques and parameters, as a primary or supplemental process in joining metallic heat shield materials for shuttle orbiter reentry vehicles. The program would also develop manufacturing capabilities and technology for vacuum-furnace-brazing all materials under consider-

ation. Concurrently, data useful to engineering and design for heat shield applications would be accumulated for dissemination. This paper also describes the fabrication of aluminum-boron metal matrix composite structures currently being considered for use in shuttle orbiter reentry vehicles.

Heat Shield Structures

Current configurations for a reusable heat shield require that large panels or sections be made from numerous smaller panels joined at their extremities, thereby forming a floating-panel concept. This concept was simulated by fabricating a test structure from Haynes 188 material approximately 6 by 36 by 72 in. The outer 3 in. of the 6-in. dimension represents the thermal protection system (TPS) portion of a multi-use reentry heat shield. The inner 3 in. of the structure simulates a vehicle inner structure.

The prime details or assemblies of the structure with which this paper is concerned consists of nine removable, open-faced, corrugated panels, 16 pi-strap assemblies, and four cruciform-strap assemblies, all suspended from the inner vehicle structure. Each of the nine panels consists of a 0.020 in. thick bead-stiffened face sheet brazed to a formed corrugation $1/2$ in. deep by 0.44 in. flat, by 1.137 in. in pitch, by 0.20 in. thick.

Figure 1 illustrates the formed corrugation. Each pi-strap assembly consists of two formed angles 0.020 in. thick and a base plate 0.030 in. thick by 2 in. wide. Figure 2 illustrates the detail for the pi-strap assembly. The cruciform consists of six formed angles 0.020 in. thick and a baseplate 0.030 inches thick by 6 inches square. Figure 3 illustrates the cruciform details.

In all, 29 separate assemblies were brazed. Of the nine corrugated panels, two were approximately 18 by 18 in., four were 9 by 18 in., two were 9 by 36 in., and one was 18 by 36 in. In addition to the Haynes 188 panels, two 18 by 36 in. corrugated panels of Inconel 625 were brazed. Each panel consisted of a bead-stiffened face sheet and a corrugation and had the same material dimensions as the Haynes 188.

Brazing of Haynes 188 and Inconel 625

Before the structural assemblies were brazed, test specimens were brazed to determine the joint strength of the brazing material. Single lap-shear brazed specimens were used. For comparison and evaluation, tensile tests were included. In both types of tests, the material was 0.020 in. thick Haynes 188.

Three 6 by 12 in. lap-shear panels were brazed, and test specimens were taken from these panels. The panels, each representing a different cleaning process, were fabricated by brazing two pieces of 3 by 12 in. sheet along their 12 in. lengths. Some of the Haynes 188 structural details were to be chemically milled to obtain the required material thickness. Therefore, the effect of chemical milling on the brazing process was to be determined in these tests.

One panel was made from as-received sheet that was only acetone-cleaned before brazing. The second panel was made from as-received sheet that was chemically cleaned before brazing. The third panel was made from material that was chemically milled first, then chemically cleaned before brazing. The cleaning procedure used is shown in Table 1.

The joint gap of the lap-shear panels was maintained at 0.003 in. along its full length. Small tabs of PH-15-7 molybdenum foil were located at five locations in the joint to assure a constant 0.003 in. gap. The overlap distance of each joint was 0.100 in. (5t—five times the 0.020 in. material thickness). The powdered brazing filler metal used on each panel conformed to AWS Class BNi 5. The braze material was held at the joint area by a clear acrylic binder. This joint gap and overlap configu-

Fig. 4—Haynes 188 vacuum-furnace-brazed pi-strap

Fig. 5—Haynes 188 vacuum-furnace-brazed cruciform

Table 1—Cleaning Procedure for Haynes 188 and Inconel 625

Operation	Constituents	Remarks
1. Degreasing	MEK or trichloroethane	Hand-rub or vapor suspension
2. Alkaline cleaning	Turco HTC or equivalent, deionized water[a]	5–10 minutes at 165–210° F
3. Rinse immersion	Water, tap or deionized	Room temperature
4. Acid pickle immersion	Nitric acid (42° Bé),[b] hydroflouric acid (70%),[c] deionized water[d]	5–10 minutes; room temperature
5. Rinse	Tap water	Room temperature
6. Smut removal	Nitric acid, balance water deionized[e]	20–30 minutes; room temperature
7. Rinse, immersion, or spray	Deionized water	Room temperature
8. Dry	Air or oven	Until dry, 200° F maximum

[a] 6-8 oz/gal.
[b] 3 parts by volume.
[c] 2 parts by volume.
[d] 2 parts by volume.
[e] 40–60% by volume.

Table 2—Haynes 188 Brazed Joint Tests—Lap-Shear Data From Room-Temperature Tests

Specimen number	Chemical processing[a]	Width, in.	Lap, in.	Lap area, in.²	Ultimate shear stress, ksi	Failure mode
2A-1		1.007	0.100	0.1007	14.55	Braze joint failure
2A-2	A	1.010	0.100	0.1010	14.85	Braze joint failure
2A-3		1.010	0.100	0.1010	14.23	Braze joint failure
2B-1		1.001	0.090	0.0901	13.04	Braze joint failure
2B-2	B	1.002	0.090	0.0902	13.77	Braze joint failure
2B-3		1.002	0.090	0.0902	14.42	Braze joint failure
2CC-1		1.003	0.090	0.0903	12.86	Braze joint failure
2CC-2	C	1.004	0.100	0.1004	13.27	Braze joint failure
2CC-3		1.004	0.100	0.1004	16.77	Braze joint failure

[a] A—acetone cleaned; B—chemically cleaned; C—chemically milled and chemically cleaned.

ration was calculated to force failure at the braze joint during subsequent mechanical testing, thereby determining the actual brazing filler metal strength of this specific single-overlap configuration.

Brazing was performed in a furnace with a vacuum better than 1 μ and at temperatures of 2150 to 2175° F for 10 minutes. Then the material was slow-cooled. After the brazing, 1 by 6 inch lap-shear-type specimens were machined for each panel.

The lap-shear and tensile specimens were tested at room temperature and at 1800° F. Tables 2 and 3 list the lap-shear data of brazed specimens tested at room temperature and at 1800° F, respectively. As indi-

cated, the pre-braze chemical surface treatment showed little or no effect on braze properties at room temperature and no discernible effect at 1800° F. Results show that the average ultimate shear stress is 14.19 ksi at room temperature and 7.24 ksi at 1800° F. Table 4 lists the tensile test data of Haynes 188 base metal and the effect of chemical surface treatments and thermal processing when specimens are tested at room temperature and at 1800° F.

As shown, the average ultimate stress of chemically cleaned specimens tested at room temperature (no thermal processing) was 139 ksi. Chemically milled and chemically cleaned specimens were slightly lower, averaging 137 ksi. Significantly, however, comparably cleaned tensile specimens subjected to a brazing temperature of 2150 to 2175° F prior to room temperature testing showed a marked reduction in tensile strength of approximately 15%. In addition, tensile specimens having chemically milled and cleaned surfaces and subjected to temperatures of 2150 to 2175° F again showed slightly lower average tensile properties at 1800° F than did specimens that were only chemically cleaned.

The Haynes 188 and Inconel 625 structural panels, pi-straps, and cruciforms were chemically cleaned, as shown in Table 1. After the assemblies were cleaned, the powdered brazing filler metal conforming to AWS Class BNi 5 was applied to the joint interface with the aid of an acrylic binder. Type 304L stainless steel bars were placed on the assemblies to maintain pressure at the mating surfaces. Brazing was performed in a furnace having a vacuum less than 1 micron at temperatures of 2150 to 2175° F. The temperature was held for 10 minutes, and the materials were slow-cooled. Figures 4 through 7 are photographs of the brazed materials. Figure 4 is the pi-strap assembly. Figure 5 is the cruciform. Figure 6 is the corrugated side of the panel brazed to the bead-stiffened face sheet. Figure 7 is a close-up of the corrugated panel.

Figure 8 is a photograph of the entire test structure, with brazed panels, pi-straps, and cruciforms assembled. This concept of the reusable heat shield panel section is currently undergoing exhaustive thermal and cyclic testing. It may become one of the major designs for future thermal protective system panels.

Brazing of Columbium and TD-Ni-Cr

Brazing development programs are in progress for

Fig. 6—Vacuum-furnace-brazed Haynes 188 corrugated panel with bead-stiffened face sheet

Fig. 7—Close-up of Haynes 188 corrugation brazed to face sheet

Table 3—Haynes 188 Brazed Joint Tests—Lap-Shear Data From Tests at 1800° F

Specimen number	Processing[a]	Width, in.	Lap, in.	Lap area, in.²	Ultimate shear stress, ksi	Failure mode
2A-4		0.981	0.100	0.0981	7.34	Base metal failure
2A-5	A	0.984	0.100	0.0984	7.09	Base metal failure
2A-6		0.987	0.100	0.0987	7.32	Base metal failure
2B-4		0.977	0.100	0.0977	7.32	Braze joint failure
2B-5	B	0.962	0.100	0.0962	7.12	Braze joint failure
2B-6		0.979	0.100	0.0979	7.48	Base metal failure
2CC-4		0.974	0.100	0.0974	7.37	Base metal failure
2CC-5	C	0.937	0.100	0.0937	7.08	Braze joint failure
2CC-6		0.963	0.100	0.0963	7.05	Base metal failure

[a] A—acetone cleaned; B—chemically cleaned; C—chemically milled and chemically cleaned.

Fig. 8—Haynes 188 thermal protection system test structure with brazed panels, pi-strap, and cruciform

the columbium alloys C-129Y and Cb-752 and the thoria-dispersed nickel-chromium (TD-Ni-Cr) alloy. Among the specific areas being investigated are brazing alloys, techniques, vacuum pressures, brazing temperatures, contamination, tooling, and brazing insulations. A major phase of the effort will be devoted to columbium coating materials and their techniques, as well as their compatibility with brazing filler metals.

Brazed Aluminum-Boron Metal Matrix Composite Structures

The Space Division of North American Rockwell has been pursuing development of boron-reinforced composite materials. The high strength, high modulus of elasticity, and low density of boron generates considerable interest in composite materials for future high-efficiency space vehicles.

Metal matrix composites reinforced with boron are being actively evaluated and developed. Previous work

with such composites was almost exclusively concerned with boron-reinforced diffusion-bonded aluminum sheet. Why aluminum? Aluminum is a logical choice for a matrix material. It has a low density, fair temperature capabilities, and good corrosion resistance to atmospheric environments. However, aluminum-sheet-boron-reinforced composites do have shortcomings. Aluminum forms a tenacious stable oxide, which makes subsequent diffusion bonding difficult. Equally important, the use of pre-bonded sheet composite panels is restricted, because only flat sheet composite panels have been available, not curved or formed structures.

Concepts for creep-forming pre-bonded metal matrix composite materials have led to some interesting and useful results. However, the creep-forming process and required technologies are highly sophisticated, and the process is, therefore, not widely used. The shortcomings of this forming approach have been recognized by numerous organizations active in this field. This has led to considerable commercial activity in producing aluminum-boron single-filament-layer (monolayer) tapes.

Aluminum-Boron Tapes

The aluminum-boron-tape concept consists of spaced parallel boron filaments backed by aluminum foil and plasma-sprayed with aluminum alloy. Total thickness of the tape is approximately 0.010 in. The tapes are commercially available with different foil backings suitable for use with either the diffusion bonding or brazing processes. The tapes can be staggered to make splices, thereby effectively removing the requirements to produce large sheets or joining smaller ones together.

An important attribute of aluminum-boron tape is its simplicity in lay-up for final forming during brazing or diffusion bonding. In addition, unidirectional, cross-ply, or multi-directional filament orientation can be achieved during tape lay-up. Tapes can be laid up on contoured fixtures, with the desired filament orientation used for each ply, with a sufficient number of plies to yield the required design thickness. The tape layers can be joined by either diffusion bonding or brazing to obtain the final formed product.

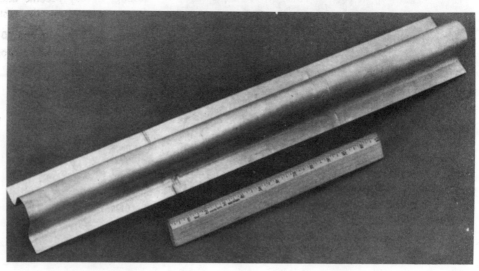

Fig. 9—26 in. contoured composite stringer assembly using aluminum-boron tapes vacuum-brazed in isostatic gas-pressure furnace

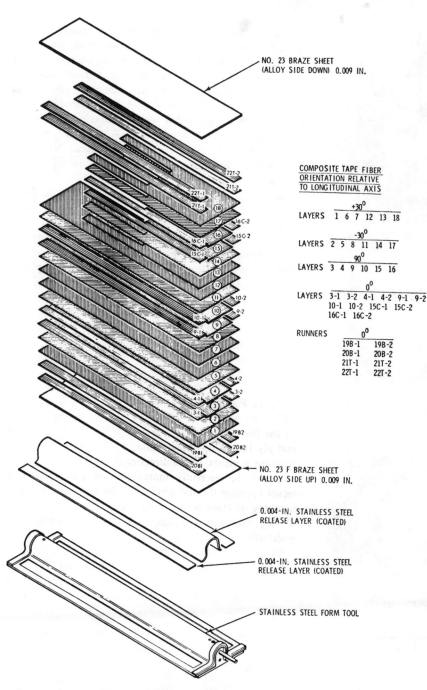

NO. 23 BRAZE SHEET
(ALLOY SIDE DOWN) 0.009 IN.

COMPOSITE TAPE FIBER
ORIENTATION RELATIVE
TO LONGITUDINAL AXIS

+30°
LAYERS 1 6 7 12 13 18

-30°
LAYERS 2 5 8 11 14 17

90°
LAYERS 3 4 9 10 15 16

0°
LAYERS 3-1 3-2 4-1 4-2 9-1 9-2
10-1 10-2 15C-1 15C-2
16C-1 16C-2

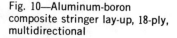

RUNNERS 0°
19B-1 19B-2
20B-1 20B-2
21T-1 21T-2
22T-1 22T-2

Fig. 10—Aluminum-boron
composite stringer lay-up, 18-ply,
multidirectional

NO. 23 F BRAZE SHEET
(ALLOY SIDE UP) 0.009 IN.

0.004-IN. STAINLESS STEEL
RELEASE LAYER (COATED)

0.004-IN. STAINLESS STEEL
RELEASE LAYER (COATED)

STAINLESS STEEL FORM TOOL

Recent fabrication of a contoured, brazed, composite structure utilizing the aluminum-boron tape is being seriously reviewed for possible application to shuttle vehicles. Generally, this completed structure (Fig. 9) can be described as a 26 in. long stringer or hat section approximately 3 in. wide. The first 17 in. of the hat section consists of 12 plies or layers of tape. The last 9 in. length is drastically increased to 18 layers. Conversely, however, to complicate the assembly and fabrication, the legs or brim of the section are radically changed in design. The first 9 in. of the brim is 16 plies, while the remaining 17 in. length suddenly is increased to 22 plies. To complicate the assembly further, boron filament orientation relative to the longitudinal axis is multi-directional—that is, +30, −30, 90, and 0 (unidirectional) deg.

Fabrication

How is this structure fabricated and what is required? First, the aluminum-boron tape containing the substrate aluminum alloy foil for brazing applications is used. Also, for a structure having a final formed configuration to be obtained, a contoured fixture precalculated to yield the final design configuration and dimensions is necessary.

Figure 10 illustrates the aluminum-boron composite stringer with filament orientation and assembly lay-up in sequence. The contoured stainless steel fixture actually controls the final configuration of the part. The fixture is designed so that it also becomes the base of the vacuum retort to be used. The first detail assembled on the fixture is a preformed stainless steel foil that acts as a release layer between the composite structure and the

Fig. 11—Close-up of aluminum-boron tapes at transition area from 12 to 18 layers

Fig. 12—Composite details shown between contoured retort base and outer retort skin before closure

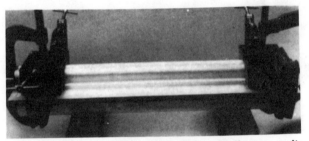

Fig. 13—Outer retort skin drawn down over the composite details prior to welding of retort peripheral flange

contoured fixture. This layer is also coated with a release agent to ensure complete release of the part.

The first detail of the actual part to be laid up is a brazing sheet. Brazing sheet, used on this structure for cosmetic purposes only, is an aluminum alloy sheet having a brazing filler metal clad on one surface. This brazing sheet is assembled with the brazing filler metal side facing the composite tapes. It actually becomes the inner surface of the completed structure. Next, the aluminum-boron tapes are cut to the required dimensions and configurations.

Tape Cutting. Accurate tape cutting is essential for a close fit between matching tapes of a single ply or layer having a change of filament orientation at specific areas of the ply. The required configuration of each ply is first put on Mylar sheets. When these configurations are cut from the Mylar sheet, they become accurate templates. The templates are then used for cutting each ply of aluminum-boron tape. Since the dimension of each ply varies slightly, a Mylar template must be made for each piece of tape in each ply. This composite stringer assembly consisted of 50 separate pieces of tape.

The tapes are then immersed in an acetone vehicle, ultrasonically cleaned, and allowed to dry.

Tape Lay-Up. Tapes are laid up in the sequence shown in Fig. 10. Note that the filament orientation is +30, −30, 90, 90, −30, +30 deg across the hat section for the first six plies. This sequence is repeated until the final ply is laid up. Meanwhile, the legs or brim of the structure consists of unidirectional (0-degree) runners and must be carefully matched with their adjoining details forming the hat section of the structure. Careful matching of tapes is also required in layers that contain 90 deg filament orientation at the hat section and unidirectional orientation at the brim.

As each succeeding tape layer is assembled, the tape is tack-welded to the previous tape. A capacitor discharge-type spot welding unit is used for the welding. Figure 11 is a photograph of the assembled structure near

Table 4—Haynes 188 Alloy Tensile Test Data

Specimen number	Chemical processing[a]	Thermal processing[b]	Thickness, in.	Width, in.	Area, in.²	Yield stress, ksi	Ultimate stress, ksi	Elongation, in 2 in., %	Young's modulus "E", psi × 10⁶
Tested at room temperature:									
1A-1		None	0.0190	0.500	0.0095	71.58	139.15	58	31.2
1A-2	B	None	0.0190	0.500	0.0095	—	140.21	60	—
1A-8		Yes[b]	0.0195	0.501	0.0098	48.61	119.38	62	31.6
1B-1		None	0.0218	0.503	0.0110	62.27	138.18	46	28.4
1B-2	C	None	0.0216	0.502	0.0108	62.96	136.57	53	26.9
1B-8		Yes[b]	0.0202	0.501	0.0102	52.48	113.86	32	28.8
Tested at 1800° F:									
1A-4			0.0190	0.497	0.0094	22.78	41.38	16	13.4
1A-5	B	Yes[b]	0.0190	0.500	0.0095	18.42	30.00	21	13.0
1A-7			0.0190	0.501	0.0095	17.68	29.47	29	11.4
1B-4			0.0202	0.498	0.0101	19.31	34.16	29	10.8
1B-5	C	Yes[b]	0.0202	0.500	0.0102	19.80	32.67	27	11.8
1B-7			0.0202	0.498	0.0101	19.50	32.50	33	8.4

[a] B—chemically cleaned; C—chemically milled and chemically cleaned.
[b] Subjected to 2150–2175° F braze temperature.

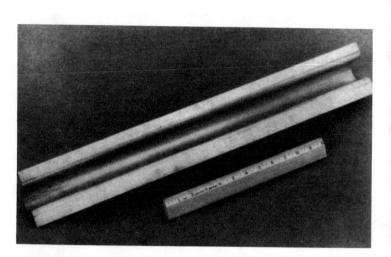

Fig. 14—Inner side of brazed contoured composite stringer assembly as removed from retort

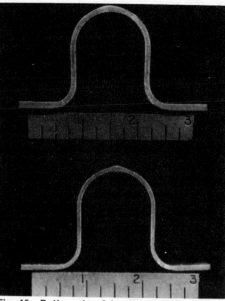

Fig. 15—Both ends of brazed stringer. Upper view shows 18 plies of composite tape at hat section; lower view shows 12 plies of tape

completion; it shows the transition area from 12 to 18 layers. Note that the transition from 12 to 18 layers is effected by a gradual change in length of each of the last six plies, rather than by an abrupt or sudden change in thickness. Figure 11 also shows the small spot tacks used to attach each tape detail. The transition area shows the adjoining matched edges of the 15th and 16th layers, where tho two plies of 90 deg filament tapes are matched with their respective runners containing the unidirectional filaments.

As lay-up of the remaining details is completed, a dummy filler assembly consisting of a 6-ply thickness is assembled on the 12th ply. This brings the top of the composite thickness equal along its full 26-inch length. The final detail of the sequence lay-up is the top or outer cosmetic brazing sheet. It is assembled with the brazing filler metal against the last layer of composite tape.

The Composite Structure. Figure 12 is a photograph showing the composite structure covered with a stainless steel sheet. This sheet constitutes the outer skin of the retort. It has been preformed to facilitate final closure. Figure 13 shows the outer skin of the retort drawn down over the composite structure and held in place with contoured blocks. The stainless steel sheet is fusion-welded to the peripheral flange of the contoured fixture. The composite structure is therefore completely enclosed in a steel retort formed by the contoured fixture and outer stainless steel sheet. The retort is then evacuated of air and sealed off by a valve in the evacuation line until the bake-out operation begins.

The retort and the composite structure within are subjected to an 800° F bake-out. This is done to assure that all contaminants that might adversely affect the brazing process have been removed. During the bake-out, the vacuum valve is opened, and the retort is continuously under evacuation. After a 30 minute bake-out, the retort and its part are cooled, the evacuation line is sealed a few inches from the retort, and the remaining line and valve are removed from the system.

Brazing Procedure and Results. Brazing is performed in an electric isostatic gas-pressure furnace. When the assembly is heated to 1000° F, argon gas pressure of 500 psi is introduced into the furnace. The gas pressure exerted against the thin steel retort cover sheet molds the composite tapes against the contoured fixture. Heating is continued to 1100° F and held for 10 minutes. Then the power is shut off. When the part cools to below 1000° F, the argon gas pressure is vented, and the part is allowed to cool to room temperature. Temperature of the part is determined by thermocouples imbedded in the contoured steel base of the retort.

Figure 14 is a photograph of the inner side of the brazed composite structure as removed from its retort. Impressions of the edges of the unidirectional runners on the brim of the assembly are visible through the inner cosmetic aluminum sheet. Figure 15 shows the two ends of the brazed structure. The upper view is the thicker section of the part. The hat section and lower radii consist of 18 plies of composite tapes, while the legs or brim are 22 plies. The lower view shows the thinner section of the part. Here the contrast is quite evident. The hat section and lower radii consist of 12 plies of composite tape, while the brim has 16 plies.

The ability to braze, in vacuum, contoured assemblies containing multidirectional oriented boron filaments in an aluminum matrix paves the way for a whole new concept in fabrication of contoured composite structures. This concept and the use of contoured composite structures fabricated from tapes are currently being reviewed for use on shuttle vehicles. It is expected that the uses of metal-matrix composite structures on shuttle vehicles will be numerous, both with respect to design configurations as well as the number of assemblies to be fabricated.

Acknowledgement

The author wishes to express his appreciation to the following persons for their valuable work and guidance in support of these programs: W. C. Harmon, D. D. Helman, and M. J. Mitchell, senior research engineers, and R. M. Heisman, supervisor. All are associated with the Advanced Manufacturing Technology and Development group of the Space Division, North American Rockwell.

Reprinted from *Tooling & Production*, January 1984

Drilling
superalloy parts for the space shuttle

With the exception of some refractory metals, Ni-base superalloys are probably the least machinable of the high-strength thermal-resistant materials. Strain hardening, the generation of high heat during cutting, welding to the cutting tools, and high shear strength combine to offer serious machining problems. These materials are also abrasive and chemically reactive.

This was the situation faced by engineers at United Space Boosters Inc (USBI), a subsidiary of United Technologies, Huntsville, AL, when they undertook the machining of a group of Inconel 718 parts for the NASA space-shuttle program.

One part, a large hex nut that is used to hold the shuttle and booster rockets in position on the launch pad, proved exceptionally challenging. It is 5½" long and requires drilling a 3"-dia hole that is later threaded.

The drilling operation is performed on a 20-year-old Warner & Swasey Model 3A, 15-hp heavy-duty turret lathe. According to Bill Perry, machine-shop foreman at USBI, most of the difficulty encountered in machining the Inconel 718 material is a result of tremendous tool pressures required because the material simply does not want to separate.

Initially, Perry tried a variety of conventional tools in an attempt to drill the holes. The problem was slow machining rates and poor tool life.

"We tried premium high-speed steel drills," says Perry, "but the material was too tough for these tools. Even with a 10-percent cobalt drill, we were lucky to get ¼" into the part before the tool needed resharpening. And if we waited too long before stopping to resharpen the drill, the material would work harden and force us to scrap the part."

Results of tests with conventional carbide drills proved equally disappointing. The cutting edges of the carbide tool wore quickly, causing work hardening of the part that led to premature tool failure.

At this point, Perry decided to try a 3"-dia CarboDrill™ indexable insert drilling bar from Carboloy Systems Dept, General Electric Co. The cutting end of the drill has two flutes, each of which has one or more recessed pockets to locate the indexable inserts.

The inserts are held in place with locking pins. The cutting edge across the diameter of the drill is divided into segments using three inserts in the larger sizes above 2⅜" dia. Two inserts are used in the smaller sizes.

One insert is placed adjacent to the centerline and carves out an annular ring with a very small ID. The other inserts are positioned farther out radially to cut the remaining annular ring.

Using the drilling bar, the operation is performed at 93 rpm and 0.0075-ipr feed, with a speed ranging from zero at the center cutting edge up to 74 sfm at the outside of the hole. Now, two of the 3"-dia × 5½"-long parts can be completed before the inserts must be indexed.

Perry attributes the CarboDrill's success in handling the tenacious Inconel 718 material to the use of Grade 390 inserts for the triangular (inboard) center cutting edges and titanium-carbide-

CarboDrill indexable-insert drilling bar. Dimension "D" shows drill diameter; dimension "A" indicates maximum drill depth (up to 7½" in 3"-dia drill size). Note coolant ports inside shank of drill.

Wear-resistant coating and specially designed chip-groove geometry of Carbo-Drill indexable drilling bar provide significant increase in productivity and quality improvement in machining Inconel 718 parts at United Space Boosters Inc, Huntsville, AL. Pictured are, left to right: Bill Perry, supervisor; Bob Allender, machine operator; and Jake Cole, sales engineer for Carboloy Systems Dept.

coated ProMax™ 518 inserts for the square (outboard) cutting edges.

"In drilling from solid," says Perry, "the cutting pressures are concentrated at the center of the drill. So, it's vital that the center cutting edges remain sharp. Grade 390 inserts have the toughness to stand up to the requirements of center machining.

"Since coated carbide is much less prone to built-up edge, the Grade 518 inserts in the outboard position are more resistant to welding with the workpiece material. Together, these inserts provide improved drilling performance that no single insert grade can."

Specially designed, advance chip-groove geometry (−55 degrees) used in the drill is also an integral element in the performance of the total drilling system. The −55-degree chip groove enables force reduction and efficient chip breaking, helping ease the chip pressure at the cutting edge and aiding chip removal.

Consequently, the superior performance of the total drilling system is due to the combination of the drill design and associated use of the coated grade-geometry system. Based on the results of this successful application, USBI now uses the system in all Inconel 718 drilling applications related to the space-shuttle program.

Seven tips to better productivity with indexable-insert drilling bars

Dramatic results have been achieved in drilling with indexable insert bars. They can often provide an 8:1 or better productivity improvement compared to HSS drills, while eliminating the need for resharpening these drills.

Common to all successful applications has been careful selection of application areas and properly established operating conditions.

The following tips for applying CarboDrill bars were provided by engineers at Carboloy Systems Dept, General Electric Co.

1. Use as short a drill as possible for your application. Under most circumstances the maximum length-to-diameter ratio should be limited to 2.5:1.

2. Adequate horsepower must be available. Metal removal rate on a 3″ drill running at 500 rpm (400 sfm) and 0.010 ipr is 35 cu ipm. Selection of operating conditions is based on the diameter of the drilling bar and workpiece material. A chart listing approximate horsepower requirements and suggested feeds and speeds for a variety of materials is available from Carboloy Systems Dept-MD130, General Electric Co, Box 237, GPO, Detroit, MI 48323.

3. Adequate thrust. Most machine tools are not rated for maximum spindle thrust. If there is inadequate capacity, the drill will not penetrate the workpiece at the set feed rate. To alleviate this, reduce feed rate.

4. Use a rigid setup. A precise set-up cannot be accomplished on a machine that lacks rigidity. Check to be sure that the spindle and gibs are tight. The turret should also be checked for rigidity.

5. Use an adequate coolant flow. Coolant aids in protecting the cutting edge from buildup and excessive heat, and serves to flush the chips back through the flutes. Although successful drilling has been accomplished with minimum coolant flow, it is recommended that a minimum of 20 psi and 0.5 gpm per 1″ drill diameter be used.

6. Center the drill. Field tests have indicated that proper centering is critical to successful application. Don't assume that the drill is on center.

7. Drill from solid. No predrilled starter hole is required. Angular entry up to 5 degrees is possible on many softer materials; however, on extremely hard materials a square face is recommended. Index the inserts when required to maintain sharp cutting edges.

Reprinted from *Machinery and Production Engineering*, June 24, 1981

Rollforming That's Out Of This World

by Brian Kellock

With the successful flight of the Space Shuttle, the possibilities for manufacture in space took an enormous step towards reality. And one of the projects for which the space laboratory will be used is the demonstration of a machine which will roll form and fabricate rigid, low-density latticework structures from carbon fibre composites. The machine is an automated beam builder, designed and built by the Grumman Aerospace Corporation for the NASA Marshall Space Flight Center.

Machines like it will eventually be fabricating structures too large to be transported into orbit from earth. According to scientists at Grumman, the fabrication of such structures, sometimes thousands of feet long, could revolutionize life. Not only will gigantic communication satellites be feasible, and solar energy stations that beam the sun's energy down to earth, but it will also enable the establishment of factories where manufacture and research can be carried out in a weightless environment.

It was envisaged that the beam sections would be formed from flat aluminum strip, stored in coils on the side of the machine. The first Grumman unit was designed to handle aluminum in this form. But carbon fibre composites were also considered because they offer greater strength and rigidity, as well as a lower density with minimal distortion throughout a range of temperatures. The problem with composites, in which the carbon fibre reinforcement is bonded with thermosetting resins, is that once the bonding sets the material cannot be reformed.

This means that finished-formed beam sections would have to be transported into space for fabrication, or the composite material would be manufactured aboard the space ship. The latter alternative presents problems of material handling as well as worries over contamination of equipment by solvents. The upshot was to develop carbon fibre composites in which thermoplastics resins--instead of thermosetting resins--were used for bonding. Because thermoplastics do not set irreversibly after forming, the composites can be manufactured on Earth in flat strips, stored in coiled form like aluminum, then decoiled for thermoforming on the space ship.

From the numerous thermoplastics material available for use as resins, two were selected: an acrylic, and a new high temperature plastics polyethersulphone developed by ICI under the trade name Victrex. For both composites the reinforcing material is a two-ply woven carbon fibre mat, manufactured by the British company Specmat Ltd. of Witney in Oxfordshire. To produce trial strips of the materials, the composite is press moulded into 1 by 3 m sheets of about 0-9 mm thick. This is approximately twice the thickness of the original aluminum sections to give the same weight per unit length. The moulded sheets are subsequently cut into 154 mm wide strips. Although at present only short lengths are being handled, there will eventually be a need for continuous stock of up to 300 m lengths to match the reel storage capacity of the beam builder.

Comparisons between the aluminum beam sections and the roll-formed acrylic composites showed that tensile strength had almost doubled. (The actual figures are given in the chart right). The composite offered the added advantage that after primary failure under tensile load, the woven reinforcement would still carry a post failure load. The carbon fibre acrylic composites did not, however, retain their good characteristics at the higher temperatures likely to be met by structures carrying heat generating or radiating components.

It was for this reason that Grumman turned to polyethersulphone, a plastics developed specifically for high temperature applications, and which has already been used on petrol engines and vehicle braking systems where temperatures in the region of 200degC are encountered. At these temperatures the material retains its strength and stiffness-to-weight ratio, while still having a good resistance to impact at temperatures as low as--100degC. Its failure load is more than double that of an equivalent aluminum beam section, and it has a post failure load capability which is a 30 percent improvement on the acrylic composites.

Attempts to use the two new materials in the original aluminum beam builder-- modified to include a low-speed servo drive, new roll forming tools and a heater to permit thermoforming--were unsuccessful. The sections suffered flange waviness and had a tendency to skew between rolling stations. There was also some delamination. So Grumman built a new machine and successfully produced acrylic composite beams. It then went on, with equal success, to use this machine to form the high temperature polyethersulphone composites, using a new set of parameters to cater for the higher forming temperatures.

The beam builder is, in fact, three identical machines arranged around a common axis. Each has its own coiled supply of strip composite, which is continuously formed into an open-V beam section as it passes through a seven-station rolling mill. The three sections emerge from one end of the machine parallel to each other, 1 m apart to form the longitudinal corner sections of a triangular beam. Preformed composite cross braces to link the section are supplied from cannisters attached to the machine, some at right angles to the sections, some inclined. These are induction welded to the longitudinal sections to form a rigid structure of the required length.

Several ways of joining the cross braces to the sections were evaluated before induction welding was chosen. This method uses an induction heating gun to produce good fusion welds in both acrylic and polyethersulphone composites. It has a very low power consumption, in the region of 7 W. Hot stapling and adhesive bonding are processes which can be used as back-ups. The other methods evaluated--ultrasonic welding, radio frequency welding and cold stapling--all presented problems. Ultrasonic welding demanded too much power; with RF welding there were problems with arcing. And while with cold stapling a good joint was made, the debris was unacceptable.

At present the beam builder is being used for ground demonstration, producing beam specimens for test. But a lightweight flight version of the machine is to be built for mounting aboard the Space Shuttle, and some of the beams produced in orbit will be returned to Earth for test and evaluation. Others will be tested in space for thermal, dynamic and structural integrity. Also on the stocks is a space platform that will be constructed to determine the feasibility of building factories in space. In the meantime, the composite materials developed for the project are finding wider applications in commercial fields here on Earth.

Comparative data for aluminum and composite beam

Material type identity	Baseline	Thermoplastic composite	
	Metal aluminium 2024-T3	Graphite/ acrylic	Graphite/ polyethersulfone
Thickness, in.	0.016	0.030	0.030
Area, in.^{2}	0.10	0.18	0.18
Weight, lb/ft	0.12	0.12	0.12
Test length, in.	59.05	59.05	48.0
Design ultimate, lb	433		
Failure load, lb	505	925	1025
Post failure load, lb		600	800
Tensile strength long (KSI)	47.0	64.5	52.2
Tensile modulus long (MSI)	10.5	8.5	7.0
Flexure strength long (KSI)	NA	112.9	153.1
Flexure modulus long (MSI)	NA	10.1	9.9
Flexure strength trans (KSI)	NA	56.8	74.3
Flexure modulus trans (MSI)	NA	3.9	3.3

Isothermal Shape Rolling
of Net Sections

A. G. Metcalfe, W. J. Carpenter, and F. K. Rose
SOLAR Division of International Harvester Co.

NET SHAPES IN DIFFICULT-TO-WORK METALS such as titanium and superalloys cannot be produced by conventional hot working processes, and are usually machined from forgings or extrusions. Two distinct problems have been identified as the obstacles to production of net shapes by hot working. The first is the chill on the hot metal caused by the tooling, and the second is the generation of scale and subsurface contamination so that extensive cleanup and, often, overall machining must be performed. Several isothermal processes have been introduced in recent years to overcome the limitation set by surface chill, but no practical solution has been advanced previously for the scaling problem.

A new process, called Isothermal Shape Rolling or ISR, was developed a few years ago and has provided a solution to both problems. It has been established as a manufacturing process for several structural sections under a U.S. Air Force sponsored program on Contract F33615-72-C-1217. Application to other sections such as airfoils has been demonstrated by company-sponsored programs. This paper will deal primarily with the work on structural sections.

ISOTHERMAL SHAPE ROLLING

This process is performed between rolls of refractory metal with continuous heating of the workpiece by passing current between the rolls. In some cases, the workpiece is clamped in tooling and passed under one roll to shape the metal locally between roll and tooling. The current flow from the roll into the workpiece causes both the roll and workpiece to be heated so that the problem of chill by tooling, discussed previously, is avoided. Also, contamination is negligible because of the short time of heating and because the roll excludes air from the metal surface when it is at working temperature.

The ISR process was evolved from an earlier process called continuous seam diffusion bonding (CSDB). In CSDB, two pieces of metal are forge welded together by refractory metal rolls with electrical resistance heating of rolls and the metal strips. Support tooling is used in many cases and may provide the forging dies to guide the forge welded metal to form fillets at the joint. The support tooling also serves to limit the heated metal to those sections that protrude from the tooling.

--- ABSTRACT ---

Isothermal metal working (ISR) using resistance heating of the metal being worked is a new process developed by the Solar Division of International Harvester. The process is unique and can be used effectively to roll structural as well as special shapes from various alloys including titanium, stainless steel, and superalloys. Sheet or plate stock can be rolled into Z-section stiffeners and channels with square external corners and internal fillets. Bar stock can be rolled into "I" and "T" sections in one-or two-roll passes. Airfoils for gas turbine blades and vanes can be formed with rolls profiled with the contour shapes desired.

The general characteristics of the processes are described. These include: high metal recovery (better than 90%); fine surface finish (16 rms); freedom from surface contamination; major thickness reductions per pass (better than 80%); control of microstructure; and very low energy consumption in processing. These characteristics result in major cost reductions when compared with the existing methods of manufacture.

ISR is performed in equipment similar to that used for CSDB. In some cases, the metal is held in tooling when the heating and shaping are local. In other cases, the full section of the feedstock is heated and shaped by two or more rolls.

A better appreciation of the ISR process can be gained by comparison with other hot rolling processes. Fig. 1 compares conventional hot rolling and the Battelle-heated roll process* with ISR. The feedstock is preheated in furnaces for both the conventional hot rolling and the Battelle processes, but is not heated until it enters the roll in the ISR process. This preheat causes some degree of alpha case on titanium in the conventional and Battelle processes, whereas it will be shown that the as-rolled section is free of subsurface contamination in ISR. The schematics in Fig. 1 bring out the rapid chilling of the preheated section in conventional rolling that is avoided in the isothermal processes. However, the combination of contamination in preheat and the temperature limitation with superalloy rolls have been found to limit the gage in the Battelle process.

Fig. 2 presents a comparison of roll force per inch width required to produce 25% reduction per pass from different initial gages of Ti-6A1-4V. The rapid rise in roll squeeze force with lower gage in conventional rolling results from the severe chill. In practice, the lower limit of gage is over 0.10 in because tears in the chilled surface occur more readily due to the presence of alpha case. The roll squeeze force does not rise with lower gages in the isothermal processes, but the limit on thickness is reached at 0.050 in by scaling in the preheat. The ISR process is not limited in temperature, and the low roll force results from rolling higher in the alpha-beta field than the 1550°F limit given for the superalloy rolls used. Roll flattening usually limits the minimum gage and is a factor in the Battelle-heated roll process because the elastic modulus is low for heated superalloys. In contrast, the elastic modulus of the molybdenum rolls used in the ISR process is nearly 40 million psi at rolling temperature. However, the high modulus alone is not adequate to explain rolling of foils below 0.010 in thick with 12 in diameter rolls. The explanation is believed to lie in the "thermal bulge" that offsets roll flattening. By this is meant that the local hot zone shown schematically in Fig. 1 results in local expansion to offset the elastic flattening, and the roll behaves locally as if it has a much smaller radius. Fig. 3 (inset) shows one of these molybdenum rolls in an edge-upsetting operation. The local hot zone and thermal bulge concepts can be readily appreciated.

The roll forces are much lower in ISR than in conventional rolling. For example, Fig. 2 shows that the roll force for Ti-6A1-4V is approximately one-tenth that of conventional rolling mills for the same reduction per pass. This advantage may be exploited in one of two ways. The first is that light-weight equipment may be designed, or larger reductions may be made per pass. Solar has found that both routes present individual advantages. An ISR mill for local working of feed-

*Superalloy rolls heated by radiant lamps are used in the Battelle-heated roll process. Preheating of the feedstock is required.

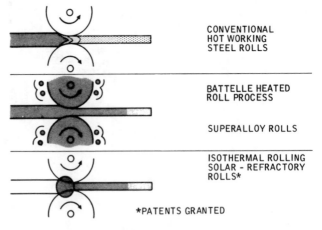

Fig. 1 - Hot rolling processes

stock, such as edge upsetting or the Square Bend process to be discussed later, can be light in construction with a maximum force capability of 20,000 lb. On the other hand, the advantage may be exploited by single pass rolling to produce thickness reductions of over 80% per pass. Such reductions are readily obtained in sections up to 1.5 in wide of Ti-6A1-4V by rolling in the research model of an ISR machine shown in Fig. 3. This machine has 40,000 lb squeeze force capability.

An important feature of the ISR process is use of feedback control to ensure constant metal temperature in rolling. In the case of thicker sections, the feedback signal is generated by a temperature sensor sighted on the metal as it emerges from the rolls. This is not possible for thin sections or sections rolled with closed rolls adjusted to roll net sections with minor flash at the points of roll contact. In such cases, some other input may be used, for example voltage drop across a shunt.

APPLICATION OF ISR TO NET SHAPES

Three types of ISR applications have been demonstrated. These are:
1. Rolling with workpiece support tooling
2. Continuous section rolling
3. Contour rolling.

Examples of each type of rolling process will be presented.

ROLLING WITH WORKPIECE SUPPORT TOOLING - The general principle in this type of rolling is to clamp the workpiece between tooling bars with part protruding beyond the tooling. The tooling and workpiece are fed through the rolls when the rolls shape the protruding metal. Side pressure is applied by small rolls as the tooling is fed through an ISR machine to ensure adequate clamping. The hydraulic cylinders for application of this side pressure can be seen in Fig. 3. Also shown in Fig. 3 is the hydraulic ram used to feed the tooling through the rolling mills.

T- and I-Section Rolling - An example of rolling with linear tooling is provided by one method to make a T-section. Fig. 4 shows the rolling of a 0.050 X 1.00 in cap on a T-section in Ti-6A1-4V in one pass from an initial section 0.30 X 0.25 in in dimension. Rolling has been interrupted and the roll raised so that the initial and final sections can be seen. Some of the

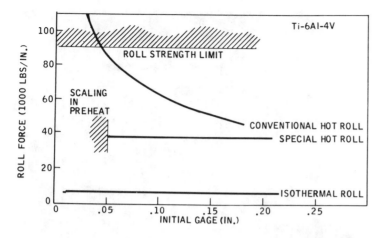

Fig. 2 - Comparison of roll force

Fig. 3 - ISR machine showing closeup of roll

Fig. 4 - Rolling of titanium alloy T-section

side rolls providing the clamping load on the linear tooling can be seen in Fig. 4. A schematic in Fig. 5 shows that this one pass rolling is the final step in the manufacture of the T-section from a piece of 0.25 in thick plate stock. Fig. 6 shows the tooling used for T-section rolling after 55 passes. The lack of wear is verified by the preservation of the machine marks on the tool surfaces.

Rolling of this T-section with linear tooling is not the most economical way to roll a T-section. This will be described under Continuous Section Rolling. However, the concept is important because it has application to other net shapes. The first is where the workpiece support tooling is not linear; for example, circular tooling may be used to generate rings with a section resembling a tee. Another important application is

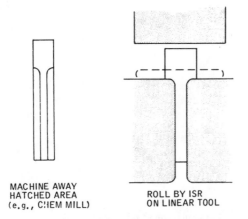

MACHINE AWAY
HATCHED AREA
(e.g., CHEM MILL)

ROLL BY ISR
ON LINEAR TOOL

Fig. 5 - Schematic of rolling method for T-section

Fig. 6 - Linear tooling after 55 passes

where the section is not constant. One example of this would be where the width of the cap of the T-section is required to be wider. In this case, the height of unreduced plate stock is increased locally so that more metal is available to be formed into a wider cap. Two passes may be required to complete the rolling of such a section. The thickness of the cap can be varied readily by cutting recesses into the top of the linear tooling.

Linear tooling has been used to make the I-beams in Ti-6A1-4V shown in Fig. 7. Again, the lowest cost method to make I-sections is by continuous rolling without linear tooling, but linear tooling offers a freedom not found with the continuous method. In addition to a programmed variation in the width and thickness of the cap (flanges), variation in height is

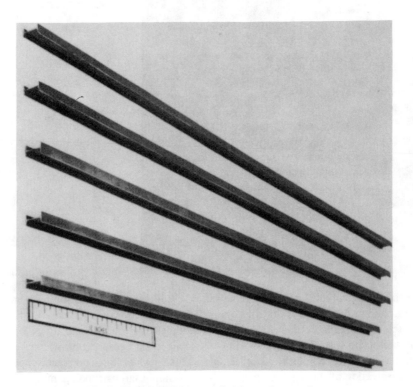

Fig. 7 - Ti-6A1-4V I-Beams rolled from plate stock

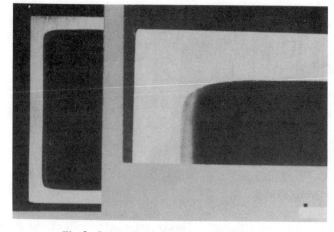

Fig. 8 - Square Bends in 0.100-in Inconel 718

(Z SECTION FORMING BY SQUARE BEND)

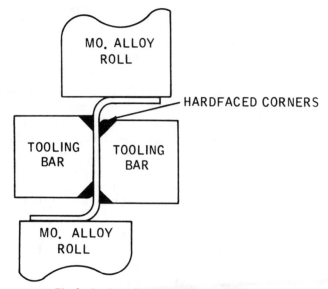

Fig. 9 - Section of progressive Square Bend

Fig. 10 - Stages in single pass Square Bend rolling

possible by use of linear tooling that would not be possible with a continuously rolled I-beam.

Square Bend Rolling - The most rewarding application of workpiece support tooling has been to produce a configuration called a square bend in sheet materials. Fig. 8 shows such a bend produced in 0.100 in Inconel 718 sheet. Noteworthy is the square external corner and the "streamlined" internal filleting. Such parts are made very readily by clamping the sheet stock between tooling bars as shown in Fig. 9. The sheet stock may be preformed by press brake operation, although this is not essential. Fig. 9 shows a setup for both bends in a Z-section simultaneously with rolls top and bottom. The sequence of operations can best be regarded as an ironing down of the cold bend to a sharp corner. Fig. 10 shows a series of sections of the initial cold bend in 0.062 in Ti-6A1-4V with successive sections as the part enters the roll until the final section gives the configuration as it emerges from the roll. The conditions for this rolling are: roll squeeze force of 5000 lb; temperature 1600-1650°F; and a speed of 10-15 in/min. The rolling temperature is attained only at the corner

that is being worked. At all other points, the Ti-6A1-4V remains cold. This has two advantages. The first is that dimensional control is excellent. For example, the height of the Z-section shown in Fig. 8 can be held within 0.005 in. In addition, the local heating permits mild steel tooling to be used, although hardfacing may be used with advantage on the corners of the tooling adjacent to the hot zone.

Fig. 11 shows three short lengths of Z-section manufactured by this process, and Fig. 12 shows a typical bend. The retention of the flow lines in the original sheet confirms the low temperatures required in this rolling.

The advantages of Z-sections formed by this process fall into three categories:

1. Reduced weight
2. High performance
3. Lower cost.

Reduced Weight - The reduced weight results from the improved structural efficiency compared with cold-formed Z-sections. A Z-section stiffener requiring 0.92 in width of faying surface for fasteners would have a 6t bend if made by cold forming and appear as shown on the left in Fig. 13. In contrast, tightening the internal radius of bends to 1/2t decreases the weight from 0.67 to 0.59 lb-ft (Fig. 13, right). This reduction in weight with bend radius follows the straight line plot in Fig. 14. On the other hand, the square bend permits a further reduction in weight because the faying surface can be

Fig. 11 - Four-ft lengths of Z-section

Fig. 12 - Typical bend in 0.06-in Ti-6Al-4V

Constant Faying Width of 0.92 inch

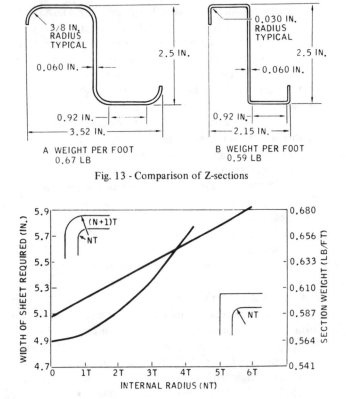

Fig. 13 - Comparison of Z-sections

Fig. 14 - Structural efficiency of Z-sections

used more efficiently. However, for other reasons, the sharp external corner is probably unacceptable for aircraft stiffeners and a compromise value of weight saving is shown by the star. Such a section has 1/3t external radii (0.020 in) and 1/2t internal radii (0.030 in), and weighs 84.5% of the conventional cold formed Z-section. It retains a 9% weight advantage over hot formed Z-sections.

High Performance - Z-sections formed by this process show high performance for two reasons. The first is that the present method of manufacture of these Z-sections is by machining extrusions so that the properties are those specified for extrusion. Sheet is the starting material in the isothermal shape rolling process and provides mechanical properties significantly higher than those of extrusions. The second reason for high performance is that the streamline at the internal fillet gives excellent fatigue strength. Flexural fatigue tests by a resonant method were performed at Solar on early material. Subsequently, additional sections made by an improved method were evaluated by Lockheed Aircraft Company. Both test programs have confirmed the excellent fatigue performance. Fig. 15 presents these results.

Lower Cost - Lower-cost results from many factors, including: excellent metal recovery, especially compared with machined extrusions; metal forming is performed in air but with negligible contamination; low-cost, mild steel tooling; nonuniform sections can be made such as variable height sections; and precision control of dimensions. The savings in strategic materials and energy compared with machining from an extrusion are enormous; these factors will become significantly more important in cost analyses in future years. At present rates, cost savings of 75% over machined extrusions are typical.

CONTINUOUS SECTION ROLLING - The general principle in section rolling is to dispense with tooling clamped to the parts and substitute rolls to guide and shape the feed stock. The rolls may be active, that is, where the roll carries current and exerts pressure on the workpiece, or passive, when the rolls act as supporting tooling. In both cases, the process becomes a continuous rolling method without the limitations set by tooling length. With increasing complexity, the roll setup may require multishaft rolling mills resembling a Turk's head mill. In many cases, it has been shown that simple roll configurations can be used to produce sections more economically than by the linear tooling approaches discussed earlier. One such application will be presented here.

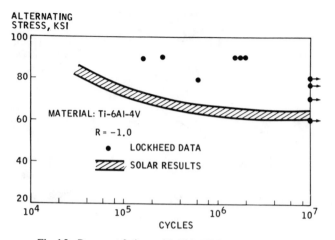

Fig. 15 - Resonant fatigue of 0.06-in Ti-6A1-4V Z-sections

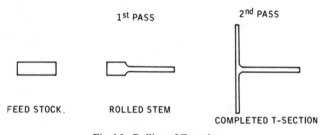

Fig. 16 - Rolling of T-section

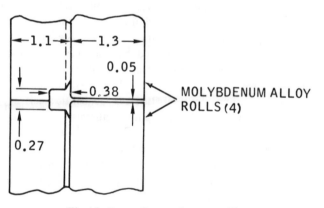

Fig. 17 - Four-roll setup for stem rolling

T-Section Rolling - The approach shown in Fig. 5 for rolling of T-sections on linear tooling has been modified for rolling T-sections from a rectangular barstock. The major decrease in height 0.30-0.05 in shown in Fig. 4 represents a reduction of 83% per pass. Such a reduction cannot be obtained in one pass in a conventional rolling process, particularly if highly finished rolls are used to produce high quality surfaces. One of the principal modifications has been to add forced feed to the rolling mill when reductions above 80% become routine. Indeed, reductions of 100% have been demonstrated; these have occurred when the feedstock has been squeezed away as the rolls have made contact leaving two opposed, feather edges. Such large reductions per pass are accompanied by major lateral flow, and it has been shown that suitable control of the force feed results in the change of section being entirely in the direction of increased width. With this redistribution of metal there is no lengthening, and precision control of the straightness of the emerging part becomes possible.

Fig. 16 shows the sequence of operations used to roll a T-section in Ti-6A1-4V in two passes from a rectangular barstock feed measuring 0.750 × 0.250 in. Fig. 17 shows the assemblage of four rolls used in the first operation. A roll squeeze force of 33,000 lb was applied with a feed force of 7,200 lb. This produced a half-dumbbell section similar to that shown in Fig. 18 in one roll pass. The flash at the head of the half-dumbbell had a thickness of 0.005-0.007 in and extended 3/16 in beyond the head. The flash at the side of the dumbbell head was thicker but extended for only a very short distance on each side.

Good dimensional control was found for the rolled stem portion. Fig. 19 shows the gage of the T-section stem for two 4 ft blanks rolled under somewhat different conditions. Both show that the 1 in deep stem may be rolled to consistent dimensions within typical sheet metal tolerances. However, the rolling conditions selected determine the length that must be rolled before consistent conditions are attained. The principal factor appears to be achievement of steady-state temperature conditions within the complex roll system.

The second pass to roll the T-section from the half-dumbbell was performed with the three-roll assemblage shown in Fig. 20. Fig. 17 shows that the height of the half-dumbbell is 0.380 in, and this is reduced in one pass to the cap of a T-section with dimensions of 0.050 in by greater than 1.50 in. This

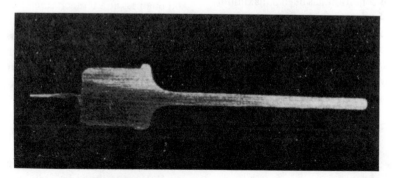

Fig. 18 - Section of half dumbbell

171

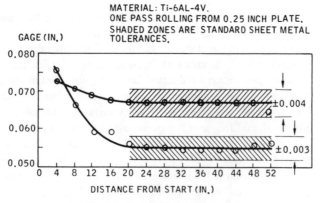

MATERIAL: Ti-6AL-4V.
ONE PASS ROLLING FROM 0.25 INCH PLATE.
SHADED ZONES ARE STANDARD SHEET METAL
TOLERANCES.

Fig. 19 - Gage of T-section stems

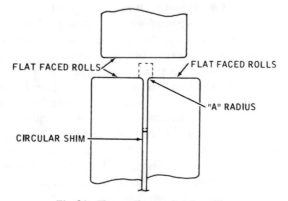

Fig. 20 - Three-roll setup for cap rolling

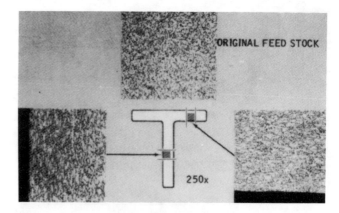

Fig. 21 - Microstructure of rolled T-section

reduction in height is 87%. Consistency of dimensions equalled those of the stems.

The rolling temperature for these T-sections had been estimated to be approximately 1700°F. Fig. 21 indicates a rolling temperature of 1670-1720°F by comparison with the alpha-beta ratios for Ti-6A1-4V provided by the work of Kellarer and Milacek (1).*

The section shown in Fig. 21 is in the as-rolled condition. The absence of alpha case and even of an increased percentage of oxygen-stabilized alpha near the surface is quite remarkable for a titanium alloy processed in air. However, the time of exposure at elevated temperature is of the order of a few seconds at the most, and during the peak temperature exposure, the metal is sealed from the atmosphere by the rolls.

As discussed earlier, control of section straightness to a very high accuracy is achieved by control of the force feed. Typical straightness in the as-rolled condition has been a maximum deviation of 0.060 in in a 4 ft length in the free state.

T-Section Costs - Materials and processing costs for a titanium alloy T-section such as the one described above have been estimated for various conditions. However, in order to make the costing and analysis of widest applicability, a simplified basis of costing has been used. This is to take purchased materials at invoice cost, add machining costs based on the U.S. Air Force Sagamore Conference data (2), and add labor

---

*Numbers in parentheses designate References at end of paper.

at $20.00/h. The cost of machining Ti-6A1-4V alloy from extrusions to T-sections is $6.20/lb according to the referenced source. Based on this approach, the cost of the 1.5 X 1.25 X 0.050 in T-section from machined extrusions is $20.00/ft. In contrast, the method of manufacture with linear tooling had been reduced to $11.10/ft, the three-roll setup method reduced the cost to $9.60 with machined blanks and to $6.30 by all-rolled sections. Similarly, the material conversion ratio is decreased from 7:1 for machined extrusions to 1:1 for the two-pass continuous rolling process.

CONTOUR ROLLING - Contour rolling includes those sections that require roll profiles other than the simple rolls used for structural sections. A good example of a contour-rolled section is provided by the airfoils used for gas turbine vane stock. However, the rolling process is similar to that used for T-sections. Force feed of round or rectangular barstock into the roll opening is an important part of the process to achieve the very large reductions required in a single pass. By application of force feed, nearly all materials can be rolled. Typical sections have been produced in metals such as: unalloyed titanium; titanium alloys; ferritic steels; austenitic steels; superalloys including Rene' 95; composite materials such as titanium-beryllium; and powder metallurgical precursors.

Simple contoured sections have been rolled to demonstrate the process. One simple section has been rolled between a flat-faced roll and a roll with a symmetrical profile consisting of a circular arc. Airfoils of 0.8 in width and a maximum thickness of 0.055 in can be rolled from barstock of 0.30-0.375 in diameter. Fig. 22 shows a simple section of this configuration rolled from 0.315 in diameter Ti-6A1-4V barstock on one pass at a temperature of 1650-1700°F. Two features are readily apparent from this figure. The first is the production of edges of 0.001 in thickness in a single pass; so sharp that as-rolled sections can be used to cut paper quite cleanly. The second feature is that the surface of the titanium alloy is completely free of any residual contamination. Also, the as-rolled surface replicates the roll surface exactly and produces finishes that may be better than 16 rms.

Fig. 23 shows the production of a simulated airfoil section in Rene' 95 alloy by this process. Again the reduction was achieved in a single pass in this high strength alloy, although leading and trailing edge radii were 0.008 in. The rolling "win-

Fig. 22 - Isothermal roll forging

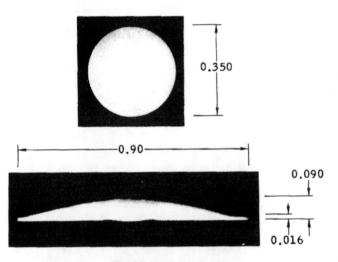

Fig. 23 - Rolled Rene' 95

dow" of temperature for this alloy is quite narrow and ranges from 2000-2050°F, but control within this range can be accomplished because of the feedback systems.

Control of metallurgical structure is another important feature of the process. For example, the alpha-beta alloy, Ti-6Al-4V, may be rolled in one of four temperature ranges, as shown in Table I.

Control of the structure is a unique feature of isothermal shape rolling and can be achieved in all roll or rolled sections in addition to the contoured airfoils. This control is much greater than in conventional working because the metal is not preheated as pointed out in Fig. 1. Preheating causes grain growth and other effects, particularly at the highest temperatures required for beta forging or rolling. In such cases, a large percentage of the subsequent reduction may be required to overcome the deleterious effects of the preheat. Unless this reduction is achieved uniformly throughout the section and at all places within the final shape, nonuniform properties will result. Such problems do not occur in beta-processed titanium alloy sections because the feedstock is not furnace preheated for long periods above the beta transus.

Further control of structure and properties is achieved by

Table 1 - Temperature Ranges for Rolling Ti-6Al-4V

Range	Temperature, °F	Product
Low alpha-beta	1550-1625	Fine alpha-beta structure
Standard alpha-beta	1650-1725	Standard structure
Beta continuous	1750-1825	Transformed islands of alpha
Beta	Above 1825	Transformed bata

heat treatments after the rolling. Heat treatment responses are particularly marked after the beta processing. Shear strength, fatigue, creep strength, and fracture toughness may be increased by as much as 20% over the standard alpha-beta processed material.

Fig. 24 compares the structure of Ti-6Al-4V barstock with the structure of isothermal shape rolled material.

Fig. 25 shows typical flash at the edge of a rolled section in Ti-6Al-4V and shows also, at higher magnification, the contamination-free and smooth-rolled surface. In the lower part of the figure, further advantage of the thin flash is shown because automatic finishing by a tumbling (barrel finishing) operation is possible. This point is brought out by Fig. 26, where conventional flash (0.010 in) and the ISR (flash 0.006 in) are compared. Not only does the automatic finishing reduce costs, but the hand grinding required for conventionally rolled parts leads to difficulty in holding the critical edge contours, chord lengths and other dimensions within the prescribed limits. Hand finishing leads to high reject rates and high inspection costs.

Several of the features that result in cost effectiveness of isothermal shape rolling to produce contoured sections, such as airfoils, have been discussed above. In summary, the key cost reduction features are:

1. Single pass rolling from barstock
2. High quality, as-rolled surface
3. Thin flash permits low cost edge finishing
4. Intermediate stage operations are eliminated (annealing, lubrication, cleanup, trimming, inspection, handling, etc.)
5. High metal recovery
6. Reduced inventory costs.

REVIEW OF ISR

The ISR processes represent new technology, and the rapid technical and economic advances expected with new technology are being found. In this regard, the experience on rolled Ti-6Al-4V T-sections on U.S. Air Force Contract F33615-72-C-1217 is illustrative. The rolling process established in late 1973 using linear tooling was analyzed to show cost reductions to half that of the persent method of manufacture from machined extrusions. The rolling process established in 1974 shows costs less than one-third those of the present method of manufacture. These costs are based on one-pass rolling of the cap and stem from barstock at 4 in/min rolling speed. There can be no question that higher rolling speeds will be realized as experience is gained. For example, partial

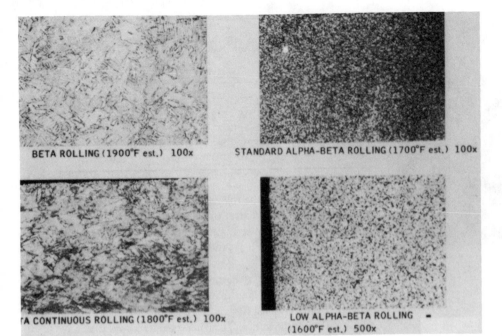

BETA ROLLING (1900°F est.) 100x   STANDARD ALPHA-BETA ROLLING (1700°F est.) 100x

TA CONTINUOUS ROLLING (1800°F est.) 100x   LOW ALPHA-BETA ROLLING (1600°F est.) 500x

Fig. 24 - Microstructure of rolled Ti-6A1-4V

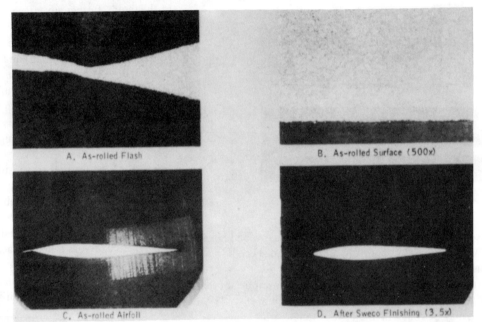

A. As-rolled Flash   B. As-rolled Surface (500x)

C. As-rolled Airfoil   D. After Sweco Finishing (3.5x)

Fig. 25 - Removal of flash

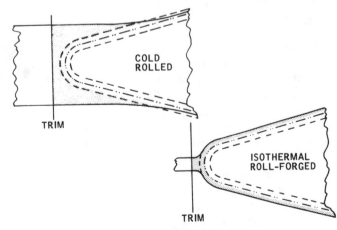

Fig. 26 - Schematics of flash

preheat just ahead of the roll entrance and higher roll force capability than is available in the present machine are expected to permit rolling at 40 in/min.

The ISR process appears to have an important role to play in reducing costs. The cost reduction comes largely from improved metal utilization. Compared with machined extrusions or ring forgings, the metal utilization may be increased fourfold, and four parts can be made by isothermal shape rolling from the same amount of metal presently used to make one part. Accompanying this advantageous use of strategic raw materials is a corresponding improvement in consumption of energy. The biggest reduction comes from the improvement in metal utilization. Metals have been described as "packaged" energy put in at the extraction, melting, remelting, heating, and reheating steps in their fabrication. More energy is expended in machining but the total energy in the chips is often completely wasted. In contrast, the high metal utilization and single heating for rolling to final dimensions represent important advantages for ISR.

The future trends are that materials and energy will rise in cost faster than labor. The ISR processes offer important advantages in all three contributors to cost, but are particularly valuable on conservation of materials and energy. For this reason, a continued growth in use of these methods can assuredly be predicted.

ACKNOWLEDGMENTS

The authors are pleased to acknowledge the support of the U.S. Air Force to apply the ISR process to square bends and structural sections, under Contract F33615-72-C-1217 monitored by Mr. Larry Clark, AFML/LTM.

REFERENCES

1. Kellerer, H. G. and Milacek, L. H., "Determination of Optimum Diffusion Welding Temperatures for Ti-6A1-4V." Welding Research Supplement, May 1970.

2. Summary of Air Force/Industry Manufacturing Cost Reduction Study, held at Sagamore Conference Center, Sagamore, New York. Report AFML-TR-LT-1, September 1972.

Reprinted from *Welding Design & Fabrication*, May 1979

King-size fuel tank boosts spacemen into orbit

*Up, up, and away! Today's supermen ride into space
powered by fuel held in huge bottles of welded aluminum*

by CARL R. WEYMUELLER, *senior editor*

Fuel tank for Space Shuttle dwarfs NASA personnel.

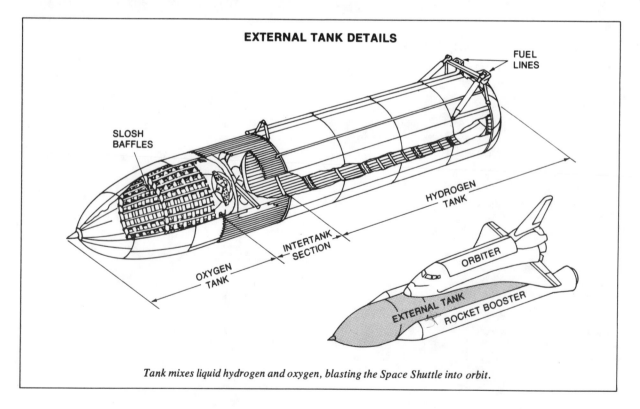

EXTERNAL TANK DETAILS

FUEL LINES

SLOSH BAFFLES

HYDROGEN TANK

INTERTANK SECTION

OXYGEN TANK

ORBITER

EXTERNAL TANK

ROCKET BOOSTER

Tank mixes liquid hydrogen and oxygen, blasting the Space Shuttle into orbit.

Looking far more like a Goodyear blimp than a fuel container, the external tank for the NASA Space Shuttle carries the liquid oxygen and liquid hydrogen that blast the orbiter vehicle into the ether. Then, its job done, the tank falls away, breaks up in the atmosphere, and splashes into the ocean where it's lost forever. Seems a shame, for a tremendous amount of costly effort and material go into each aluminum tank, which is used only once.

Built by Martin Marietta Aerospace, these tanks are fusion welded of 2219-T87, an aluminum alloy. This alloy has good weldability, plus strength and ductility from cryogenic temperatures to 600 F (320 C), needed to withstand the cold of liquid oxygen and hydrogen and the heat of aerodynamic friction during blastoff. The −T87 heat treatment, achieved by solution treating at 995 F (535 C) and aging at 325 F (163 C) for 24 hours, develops outstanding properties: 70,000 lb/in.² (480 MPa) tensile strength; 57,400 lb/in.² (395 MPa) yield strength, and 8.7-percent elongation.

Assembling the tank

All fuel tanks, six for testing and three for the flights, are built at NASA's Michoud Assembly Facility in New Orleans. Three sections make up a tank. Two fuel containers, for oxygen and hydrogen, are linked by the third, the intertank section that contains piping and instrumentation. Liquid oxygen and hydrogen flow separately to the engine through 17-inch (430 mm) diameter feed lines, where they combine violently for blastoff. Pipes empty tanks in 9 minutes.

The liquid oxygen tank, which includes the nose (ogive) section, holds about 1.33 million pounds (600,000 kg) of oxygen. Baffle plates inside keep the liquid from sloshing.

Ring moves into position for removal of barrel from weld fixture.

Domes for fuel tanks are assembled from preformed gores.

The nose has two sections, one forward and one aft. Eight preformed gores (triangular curved sheets) comprise the 15-foot-(4.6 m) long forward section. Two gores are welded to make quarter-panels. Then, these quarter-panels are welded to make halves. Finally, these halves are joined together to complete the forward ogive section.

The 17-foot-(5.2 m) aft section is also built up by the quarter-panel technique. Being larger, however, it requires 12 pre-shaped gores. Each quarter-panel, therefore, is built up of three gores.

The liquid hydrogen tank holds 224,000 pounds (102 000 kg) of hydrogen. It has a rear dome built of gores much like those for the aft section of the nose. Again, three gores, preformed, go into each quarter-section. All other sections of both tanks are essentially straight-sided barrels, short and long.

Making the welds

Tank sections are assembled by gas tungsten arc welding (GTAW), direct current, straight polarity. Square butt joints are standard, and most welds are three-pass from one side.

According to Frank Clover, the welding engineer who supervises the work, the tanks contain

KING-SIZE FUEL TANK

many thicknesses of aluminum sheet. Most welds are made out-of-position — that is, few are made flat downhand, the most convenient way, owing to the nature of job. Joint edges butt tightly, and are backed up by parallel bars separated by slight gaps that allow observers to follow the weld's progress from the back. Cover gas is 100 percent helium flowing at 100 ft³/h (2.8 m³/h); no backup gas is used.

Typical of welds in thin sheet are these parameters for joining material 0.14-inch (3.6 mm) thick. Parts are clamped in a fixture for the first pass, called a seal pass: it's made at 15 in. (380 mm) per minute; 15 volts and 75 amps. The aim is about 30-percent penetration to bring minor gaps together, creating a constant condition along the joint length for the penetration pass. For this pass, welders use filler metal, aluminum alloy 2319 in 1/16-inch-(1.6 mm) diameter wire, fed at 20 in. (510 mm) per minute. Parameters: 14 volts, 100 amps, 12 in. (300 mm)/min travel speed. The last pass, called the fill (or cover) pass, runs at 12 in. (300 mm) per minute, 14.5 volts, and 80 amps; wire feed is 30 in. (760 mm) per minute.

Weld parameters in thick sheet joined vertically are typified by those in 0.375-inch (9.5 mm) stock. The seal pass runs at 15 in. (380 mm) per minute (14 volts, 125 amps), again for 30-percent penetration. No filler wire is used for the penetration pass, which is laid in at 8 in. (200 mm) per minute; 12 volts, and 230 amps. For the fill pass, travel speed is 8 in. (200 mm) per minute, 15 volts, 180 amps. Wire feed for filler metal is 50 in. (1 270 mm) per minute.

Short (15-foot, or 4.6-m and 8-foot, or 2.4 m) barrels are assembled vertically. Longer (20-foot, or 6.1 m) barrels are made

Fuel tanks in different stages of construction.

up of eight rectangular panels of 0.34-in.-(8.6 mm) thick sheet, one panel being added and flat-welded in place at a time. Action starts when two panels are conveyed into the welding jig, and long edges are matched up for flat welds. Each weld requires three passes under 100 percent helium. The first pass runs at 15 in./min (380 mm/min), 13.8 V, 180 A; the second, at 7 in./min (180 mm/min), 12.5 V, 230 A; and the third, at 8 in./min (200 mm/min), 15 V, 195A. Filler wire feeds at 50 in./min (1 270 mm/min).

After the first two panels are joined, the section indexes ahead one panel to fit around the barrel jig. Then, the third panel is fitted to the edge of the second to be welded in place as before. As each panel moves into the jig, the finished sections roll ahead and around the jig, gradually becoming a barrel.

When completed, the barrel is clamped in a split ring to remove it from the weld fixture. Joined, these barrels build up to form the cylinders that fit on the nose shapes and domes (for liquid oxygen tanks) or domes (for liquid hydrogen tanks).

Semi-vertical welds join barrels to each other and to noses or domes. But barrels are not joined directly end-to-end. Instead, shopmen fit a narrow ring of 2219-T87 between each barrel, and two welds are laid simultaneously by a stationary GTAW machine as the barrels and ring turn on the horizontal jig. The machine deposits both welds at the same time to balance stresses; uneven stresses could distort the rings. Tack welds and backup bars hold sections in place to maintain tight tolerances throughout all welding operations.

Testing finished tanks

All completed welds are inspected by radiography and dye penetrant methods. Liquid oxygen tanks are filled with water, which is pressured to 20-22 lb/in.² (140-150 kPa), simulating calculated levels of flight stresses. Nitrogen, also under pressure, substitutes for water in liquid hydrogen tanks because water at the higher pressure (42 lb/in.², or 290 kPa) would damage the aluminum skin.

Tested tanks are cleaned and covered with a 1-inch-(25 mm) thick layer of foam (polyethylene) insulation that keeps ice from coating the tank due to the extreme cold of the liquid gases. The foam also contains a charring ablator that flakes away during launching, absorbing heat to shield the tank. A coat of paint, added for atmospheric protection, completes the tank for final assembly when all accessories are added. Testmen check out the fuel system, and the tank is ready for its maiden — and final — voyage. ■

Reprinted from *Welding Journal*, September 1980

Welding of the SPACE SHUTTLE Orbiter Crew Module

Includes the fabrication of aluminum alloy 2219 panels using the gas tungsten arc process with hard and free-fall tooling as well as aluminum alloy filler metal

BY E.L. WHIFFEN, L. J. KORB AND C. E. OLEKSIAK

The intent of the Space Shuttle mission is to provide a practical, low-cost method of transporting payloads, astronauts and scientific personnel to and from earth orbit. The system will permit the United States to orbit payloads of up to 65,000 pounds (29,545 kg) and return to earth at costs only a fraction of previous costs. It is possible only because major elements are being designed for reuse with only minimum refurbishment.

Four major flight vehicle elements comprise the Space Shuttle system; these are:

Space Shuttle Orbiter in launch position with external tank and solid rocket boosters (photograph)

Paper presented at the AWS 61st Annual Meeting held in Los Angeles, California, during April 14-18, 1980.

The authors are with Rockwell International as follows: E. L. WHIFFEN, member of Technical Staff V, Space Systems Group, Downey, California; L. J. KORB, Supervisor—Materials and Processes, Space Division, Orange, California; and C. E. OLEKSIAK, Manager of Materials & Producibility Welding Engineering, Rocketdyne Division, Canoga Park, California.

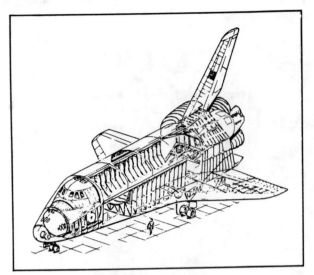

Fig. 1—Skeletal view of Space Shuttle Orbiter

1. The solid rocket boosters (SRB).
2. The Space Shuttle main engines (SSME).
3. The external tank (ET).
4. The Space Shuttle Orbiter (SSO).

The solid rocket boosters provide the major thrust during launch; they are augmented by the main engines, which burn liquid hydrogen with liquid oxygen supplied by the external tank. The orbiter, launched vertically from a piggyback position astride the external tank, contains the main engines, the astronauts, and the payloads.

The orbiter shown in Fig. 1 functions as both a spacecraft and an aircraft. During entry from each orbit, it must be protected from temperatures up to 2300°F (1260°C) on its underside and temperatures to 2700°F (1482°C) along the leading edges. At an altitude of 150,000 ft (45,720 m), the orbiter will slow to about eight times the speed of sound and will pass its maximum heating. At 50,000 ft (15,240 m), the orbiter will enter into a level flight path and will maneuver aerodynamically to land as a conventional aircraft.

The dry weight of the orbiter is limited to 150,000 pounds (68,182 kg). To achieve this weight requires the most efficient materials and joining methods. Welding plays an important role in the body structure of the orbiter, which is the largest weight element. It is used in three major areas: the crew module, the thrust structure, and the more than 40 pressure vessels required for the vehicle. The welding process controls to ensure the high reliability of the aluminum crew module are discussed below.

Crew Module Design

The crew module is shown in Fig. 2. It is approximately 17 ft (5.2 m) long and up to 18 ft (5.5 m) in diameter. The forward portion of the crew module consists of a 10 ft (3 m) diameter flat bulkhead intersecting a conically shaped body, whereas the aft portion employs both conical and double curvature panels welded to an 18 ft (5.5 m) diameter aft bulkhead. The crew module has two levels and is designed to carry a normal crew of four. A total of 10 people can be carried as is required for some missions.

Essentially, the crew module is a large aluminum pressure-tight vessel that must be capable of withstanding 16 psi (0.11 MPa) limit internal pressure with a 1.5 factor of safety and a 1.0 psi (0.007 MPa) limit collapsing pressure with a 1.4 safety factor. The latter requires that the skins be stiffened in both the longitudinal and transverse directions to prevent the shell from buckling. Because of the critical function of the crew module, its multiple mission requirements (400 missions), and its very low factor of safety (as compared with ASME code vessels), every effort must be made to ensure the highest possible quality and reliability.

Alloy Selection

The choice of the crew module structural material was quickly narrowed to the 2000-series aluminum alloys because of the combined requirements for high-strength-to-density, size and cost. While titanium is considered more efficient for pressure vessel structures, the size of the crew module precluded welding it inside

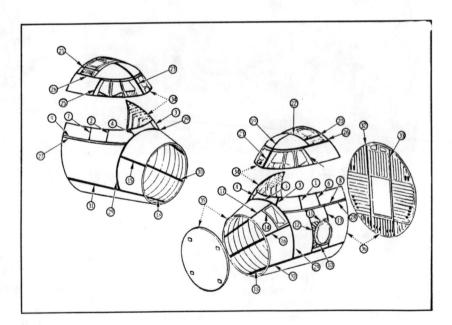

Fig. 2—Crew module weld locations

Fig. 3—Side hatch panel with integral stiffening ribs

Table 1—Alloy Selection

Characteristics desired	2219	2014	2024	2124
High strength				+
Weldability	+			
Corrosion resistance			+	+
Stress corrosion resistance	+			+
Formability (compound curvature)	(a)	+	(b)	(b)
Historical experience—SD/ industry		+		

(a) Weight penalty—cannot fully strengthen after forming.
(b) Requires ice box treatment between the solution treating and the forming operations.
(c) + indicates outstanding characteristic of the alloy.

a chamber—a control considered essential to obtaining reliable, reproducible titanium welds.

Designing for external load is economically achieved in aluminum by machining panels with integral stiffening ribs directly from thick plates—Fig. 3. Stiffening of a titanium shell is much more costly, especially where riveted skin-stringer-stiffened panels could not be permitted due to potential leakage. Because of the sizes of aluminum plates available, a minimum number of welds could be used. For example, the 18 ft (5.5 m) diameter flat aft bulkhead requires only two welds to complete its fabrication. The requirements for forming to accommodate double curvature panels, such as in the windshield areas, again favored aluminum. Thus, raw material costs, machining costs, forming costs, and welding costs all favor aluminum.

Four aluminum alloys (i.e., 2014, 2219, 2024, and 2124) were considered—Table 1. Both the Apollo command module inner cabin structure and the Saturn S-II were fabricated from 2014-T6 at the Rockwell International Space Systems Group. For this reason, it represented the alloy with the greatest experience base. Aluminum alloy 2219, however, offered the best combination of strength, stress corrosion resistance, and toughness desired for ensuring a reliable structure. Furthermore, it could be repaired by manual welding with little risk of cracking, whereas 2014-T6 does not lend itself to manual weld

repair.

Several different tempers of 2219 were used—including −T62, −T81, T851, and T87. Essentially, the T87 temper was used in panels requiring little or no forming while the −T62 temper was required for the severely formed, double contour panels, and the T851 temper for single contour panels. Although successful welding could be accomplished with either 4043 or 2319 aluminum alloy filler metal, the latter was chosen because of slightly higher properties and the avoidance of dilution of copper precipitates in areas where repair would be required.

Welding Design and Philosophy

The crew module must accommodate many penetrations, including 10 windows (six forward, two overhead, one hatch, and one into the payload bay) and three hatches (side, payload bay, and airlock), necessitating a high level of panel complexity. The basic design philosophy was to minimize the number of panels (and, consequently, the length of welding) to minimize tooling costs and increase reliability. As presently configured, the crew module consists of 36 welded members, comprising 36 welded joints totaling 4719 in. (119.86 mm) of welding. It has been designed to preclude the need for any manual welding except in isolated repair cases. All machine welds were designed as square-butt joints with thicknesses ranging from 0.125 to 0.290 in. (3.18 to 7.34 mm).

Welded joints were moved away from sharp contour changes to increase reliability. For example, both the forward and aft flat bulkheads contain integrally machined peripheral flanges to permit the joint between the bulkheads and the conical body to be made on the conical surface—approximately 4 in. (102 mm) away from the geometrical transition between a cone and the flat bulkheads.

Wherever possible, designs for intersecting welds were configured for T-intersections instead of cross-over welds. Weld lands, approximately 1 in. (25.4 mm) wide, are provided for all welded joints to compensate for the reduction of properties of heat-treated panels in the weld bead and the heat-affected zones. To accommodate different panel joint thickness requirements, certain welds had transition step areas where thicknesses increased or decreased within the weld but never at an intersection—Fig. 4.

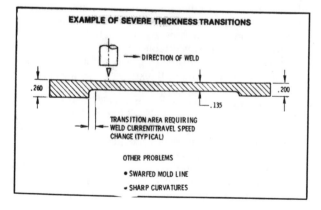

Fig. 4—Example of severe thickness transitions

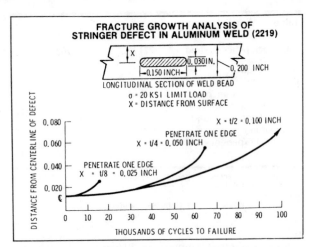

Fig. 5—Fracture growth analysis of a stringer defect in an aluminum weld (2219)

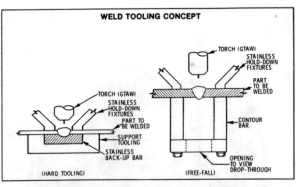

Fig. 6—Weld tooling concept (hard tooling and free-fall tooling)

Weld Strength and Criteria

Weld allowables were established using 99% probability and 95% confidence values based on actual weld test data that were derated to accommodate maximum specification deviations. Specifically, these include a maximum joint peaking of 2½ deg, a maximum offset of 10% (based upon weld centerline mismatch), undercut—lack of fill, or suckback up to 2% of weld land thickness; and a maximum transverse cross-sectional area loss of 6% due to pores or stringers.

The maximum spherical pore size permitted by specification was 1/2 t or 0.080 in. (2 mm), whichever is smaller. Pores grouped closer than the diameter of the largest pore were considered to be a single defect of the size of the circumscribing circle. Stringer-type defects up to 0.080 in. (2 mm) are permitted provided that the length-to-width ratio is not greater than two and that no sharp corners (<0.005 radius) were present.

Neither cracks nor lack of penetration nor sharp notches from suckback of undercut are permitted. Manufacturing is permitted, under the supervision of a manufacturing welding engineer, to groove out (within the weld bead only) and add one pass beyond the established weld schedule to correct any known defect condition. If the defect cannot be removed by this action, a formal material review is required involving actions acceptable to both Engineering and Quality Assurance.

All defects exceeding the specification were examined to ascertain the impact on both the static and the fatigue behaviors. The reduction of the static strength was established both by estimating the loss of the cross-sectional area, correlated with established test results of defects developed on the Saturn S-II program. Fatigue life estimates were arrived at by treating the defect as a crack. Through crack growth analyses techniques, the number of cycles was predicted until full penetration occurred for an acceptable defect—Fig. 5. This number should not exceed crew module design life, and the strength and stability of the through-thickness crack must not permit catastrophic failure of the crew module.

Manufacturing Considerations

The key to quality in the manufacture of the crew module lies in the ability to ensure accurate dimensional control of parts and assemblies and the ability to minimize welding defects.

Dimensional Control

Dimensional control begins with the machining of panels prior to heat-treating. Allowance must be made for a growth of 0.0008 to 0.0016 in./in. (0.02 to 0.04 mm) during aging. Additional trim must be provided to accommodate between 0.030 and 0.090 in. (0.76 and 2.29 mm) transverse shrinkage during welding, depending on the thickness to be welded. During welding certification, shrinkage is checked with a four-point shrink gauge on test specimens before the spacecraft panels are trimmed. Longitudinal shrinkage, roughly 0.001 in. (0.025 mm)/in., must also be considered in tooling for subassemblies.

The welding process chosen for the crew module is the gas tungsten arc welding (GTAW) process using direct current straight polarity (dcsp). This process, in combination with a 100% helium gas, produces a narrow, straight-sided bead that minimizes distortion and avoids weld peaking.

One of the key elements in achieving dimension control is the design of the tooling. The tooling must be massive and relatively rigid to maintain the necessary mold-line control during the welding operations. The tooling designed must also permit the complex sculptured panels and subassemblies to be loaded, trimmed, removed for cleaning, reloaded for welding, and then removed from the tooling after completion of the weld without the complete disassembly of the tooling.

Weld Tooling

Early in the program, an extensive Manufacturing Producibility program was conducted. The purpose of this program was to determine actual tooling and welding techniques required. Specimens of each weld joint configuration were welded and evaluated. As a result, a tooling philosophy was established to use stainless steel hold-down fingers and backing bars for all square-butt machine-welded joints in thicknesses 0.18 in. (4.6 mm) and below.

All machine-welded joints in excess of 0.18 in. (4.6 mm) thickness use a free-fall welding technique, stainless hold-down fingers, and contour backup members. The latter technique does not provide a groove in which to cast the weld nugget—Fig. 6. Penetration is controlled by a welder located on the penetration side of the weld who is equipped with an override welding control to permit him to adjust the depth of penetration. This

Fig. 7—Typical weld tool—vertical weld on windshield subassembly

technique also allows welding over varying joint thicknesses, e.g., from 0.250 to 0.150 in. (6.4 to 3.8 mm), then backup to 0.250 in. (6.4 mm) without laps or folds in the weld bead root at the various thickness transitions. These controls can also be programmed into the welding machine to perform this operation automatically.

Hold-down fingers are made from austenitic stainless steel because it is nonmagnetic and corrosion-resistant. Its low conductivity, approximately 18% of that of copper, is more forgiving of localized weld chilling than copper where contact is not perfect. The fingers have a contact area approximately 2 inches (50.8 mm) long by ¼ in. (6.35 mm) wide and are separated by a 0.030 in. (0.76 mm) nominal gap. The distance between fingers across the weld bead varies from ⁷/₁₆ to ¾ in. (11.1 to 19.1 mm), depending on the thickness. The fingers are actuated by fire hoses in the fixture using line pressures of 60 to 100 psi (0.41 to 0.69 MPa) resulting in finger pressures of up to 90 pounds (40.9 kg) per lineal inch. Fingers are designed as such that no side pressure is exerted and the parts being welded can move in a transverse direction to accommodate thermal stresses.

A total of 16 separate weld jigs were designed and fabricated for crew module assembly. Twelve of the assembly jigs were designed for subassembly-type welding operations and had from one to four separate weld stations. Four of the assembly jigs were designed for major subassembly and final closeout-type joining operations. Figures 7 and 8 illustrate examples of the weld jigs used.

One of the major tools had up to five weld stations, with a sixth station added by incorporating another welding jig. The size of assembly weld jigs can be better visualized in terms of weight. The smallest assembly jig had a weight of approximately 8000 pounds (3636 kg), and some of the major tools weighed 38,000 pounds (17,273 kg) or more. Both hard and free-fall tooling may be incorporated in multiple-station jigs. Of the total of 36 welds, 14 are welded free-fall and 22 are hard tooling.

Five additional assembly fixtures were designed for weld subassembly, frame, floor, and beam installations. All of these tools were designed to handle out-of-sequence part installations.

Finally, one of the major considerations in tool design was that of weld position. When possible, parts were fixtured for welding in the vertical or flat (downhand) positions. Horizontal position welding concepts were utilized only where tooling in other welding positions would have been too costly or impractical. Where a swarf was present along the weld axis, free-fall welding techniques were used because of the difficulty in centering the joint along backing bar grooves.

Skate Concept

The skate concept (Fig. 9) was employed to reduce the risk factor of moving large parts and the expense of elaborate handling equipment. The success of this approach was demonstrated on the Apollo and Saturn

S-II programs where expensive and massive hardware was commonplace.

The design of the crew module with complex machined (five-axis) and formed panels and many swarfed weld joints created the requirement to develop a universal skate system to allow routing, sawing, and welding in all positions except overhead. This system also provides for in-place radiography from the same skate support.

Because of the manufacturing engineering requirements for the trim to be ±0.30 in. (0.76 mm) of true position and normal to the mold line within 3 deg, and the need for matching the adjacent part to within 0.010 in. (0.25 mm), a precision skate and track system was required. The skates were designed to provide a three-point contact on the two rail tracks—two points on one rail and one on the second rail where the drive gear rack is attached. This system could track all joints to the critical dimensional requirements including the swarfed welds. A precision drive was required. This was accom-

Fig. 8—Typical weld tool—(six weld stations) major assembly welds—lower portion of crew module shown prior to loading upper assembly which contains windshield and escape panels

plished by the use of a Sciakydyne* servo drive system duplicating the travel system in the automatic fusion welding consoles. The same drive motor was used for all operations.

Welding Equipment

The Sciaky* "zero error" fusion welding unit was selected for the Shuttle crew module based upon the successful past performance on the Saturn S-II and Apollo programs, its availability, and the simplicity of modifying equipment to solid-state circuitry to increase welding control reliability.

The equipment is fully automatic and is a constant current, arc-voltage-controlled unit. It has been modified to accommodate a weld travel stepping system to effect parametric changes at weld joint thickness transitions. The master programmer in the console provides for automatic preweld and postweld sequencing parameters.

Four major parameters are controlled: travel, filler metal feed, current, and voltage. The latter two parameters are controlled to ±1%. Where free-fall tooling is used, the operator reading the root side of the bead has manual override capabilities to ensure proper penetration, and can adjust the current to ±5% and the travel feed to ±10%.

Due to the age of the Sciaky equipment, Dimetrics Gold Track fusion welding machines were phased in during the welding of the third crew module (CO 099). This equipment has complete solid-state electronics and incorporates such features as welding torch oscillation, voltage and/or weld current pulsation and excellent arc starting capabilities. This equipment performs the same control and sequencing functions as did the previous equipment.

Control of Weld Geometry

All welding tools and parameters were certified on test specimens prior to the welding of spacecraft parts. By proper tooling and the selection of proper weld settings,

Trademark of Sciaky Brothers, Inc.

the weld geometry, such as mismatch, peaking, lack of fusion, lack of fill, undercut, and concave root surface (suckback) are well-controlled. Depending on the gauge welded in the Shuttle crew module, voltage was controlled between 12½ and 15½ V, current between 120 and 180 A, travel speed between 8 and 12 ipm (3.4 and 5.1 mm/s), and the wire feeds between 10 and 40 ipm (4.2 and 16.9 mm/s). A typical heavy-section weld will have an intermittent tacking pass using 2 to 2½ long tacks spaced between 6 in. (152.4 mm) gaps. This will be followed by a continuous tack pass, penetrating approximately 30% through the weld joint. A full penetration pass is then applied without filler metal and followed by a cover pass. On thinner sections, such as 0.135 to 0.140 in. (3.4 to 3.5 mm), only three passes are used—two tacking passes and a combination penetration and cover pass.

Cracking is prevented by the use of run-off tabs for starting and stopping a weld wherever possible. When a weld must end on a part, a knob of weld metal is built up at the weld termination and machined flush to avoid leaving shrinkage cracks in the weld bead. (All weld beads are ground within 0.010 inch, i.e., 0.25 mm, of flush after all welding is completed.) This procedure avoids cracking where welds intersect by reducing localized stress concentration.

Control of Porosity

The most difficult defects to control are those of porosity. Porosity in aluminum is caused by hydrogen gas unable to escape from the weld bead during solidification. This porosity takes two forms—the conventional spherical-shaped pore and a long narrow stringer-like defect as the hydrogen bubbles follow the molten weld bead while trying to surface. Since hydrogen can occur by the breakdown of hydrocarbon of moisture in the arc, extreme care in all phases of welding control must be maintained to minimize porosity.

The control of several factors has been shown to affect porosity. There are:

- Weld position.
- Type of tooling.
- Type of process.
- Type of shielding gas.
- Manifolding of shielding gas.
- Cleanliness of filler metal.
- Cleanliness of part.
- Cleanliness of tool fixture.
- Environmental factors.
- Part fit-up.

Efforts to control these factors are discussed below.

Weld position is one of the most critical variables affecting porosity, as will be seen in the quality summary charts in Tables 2–5. The ideal welding position is the flat or vertical climb position, allowing hydrogen to bubble out of the weld bead. During horizontal welding, hydrogen rises to the upper edge of the weld nugget and is trapped by the solid metal above the weld bead. This problem is particularly accentuated by the use of the free-fall tooling. It is suspected that hard tooling permits greater heat to be put into the weld bead (because of the chilling effect of the backing bar) and allows greater boil-off of the hydrogen. Wherever possible, tooling was

Fig. 9—Weld skate shown being used for welding of side hatch subassembly

Table 2—Weld Description/Defects Repaired

Weld[a] number	Length, in.[b]	Type tooling[c]	Weld position[c]	Approx. inches of weld defects repaired[a]		
				S/C 101	S/C 102	S/C CO 099
1	27	H (FF S/C CO 099)	V	0	0	0
2	27	H (FF S/C CO 099)	V	0	0	0
3	17	HT	V	0	0	0
4	17	HT	V	0	0	0
5	105	HT	(H F S/C 101)	0	0	0
6	21	HT	V	0	0	0
7	21	HT	V	0	0	0
8	148	FF	H	0	0.44	0
9	148	FF	H	2.64	0.72	3.05
10	51	HT	F	0	0	0
11	51	HT	F	0	0	0
12	98	HT	H	0.32	0	0
13	100	HT	H	0	0	0
14	52	FF	H	0	0	0
15	63	HT	F	0	0	0
16	63	HT	F	0	0	0
17	63	HT	F	0	0	0
18	178	HT	F	0	0.51	2.72
19	72	HT	F	0	0	0
20	37	HT	F	0	0	0
21	37	HT	F	0	0	0
22	120	HT	H	0	0	0.51
23	55	HT	V	0	0	0.40
24	46	HT	V	0.11	0	0
25	46	HT	V	0	0	0
26	266	FF	H	4.62	0	0.40
27	148	FF	V	0	0	0.50
28	148	FF	V	0.37	2.36	2.00
29	320	FF	H	2.22	46.82	3.24
30	185	HT	V	0	0	0
31	185	HT	V	0	0	0.75
32	201	FF	F	0.600	0.60	0
33	201	FF	F	0	0.82	0
34	357	FF	V	2.08	5.51	4.50
35	407	FF	H	197.32	8.20	1.34
36	638	FF	H	0.88	1.76	0.36
Total	4719			211.16	67.74	19.77

[a] See Fig. 2.
[b] 1 in. = 25.4 mm.
[c] Key: HT—hard tooling; FF—free-fall tooling; V—vertical; F—flat; H—horizontal

Table 3—Welds Accepted Without Repair, %

Crew module	S/C 101	S/C 102	S/C CO 099
Total welds	36	36	36
Welds accepted	26	26	24
Welds accepted without repair, %	72	72	67

Table 4—Percent of Total Weld Length Accepted without Repair

Crew module	S/C 101	S/C 102	S/C CO 099
Length all welds, in.	4719	4719	4719
Defects requiring repair, in.	211	68	20
Weld length accepted without repair, %	95.5	98.6	99.6

designed for welding in the flat or vertical climb positions and the horizontal welding position was used only where tooling cost would have been prohibitive. Table 6 is a summary of welding made in the various positions of welding and the type of tooling employed.

Experience at Rockwell International has shown that dcsp is superior to ac welding because the bead is hotter and narrower, significantly reducing porosity as well as peaking and shrinkage. The use of helium, rather than argon or argon-helium mixtures, also increases the arc intensity and appears to be more effective in boiling off hydrogen. Helium used is procured to the Specification MIL-P-27407. It is controlled to a −78°F dew point.

The helium is stored in cylinders on the welding console and manifolded through the regulators and to the welding head through approximately 50 ft (15.2 m) of polyvinyl chloride tubing (PVC). A check valve is provided as close to the welding head as possible and lines are kept full of helium to avoid moisture pickup. Approximately 90 to 110 cfh (42.4 to 51.9 liters/min) of helium at 30 psi (0.21 MPa) are used. Prior to starting the weld, the helium gas is run through a moisture monitor directly from the torch cup. A ⅝ in. (15.9 mm) ceramic cup is used at the torch to provide inert gas shielding of the weld. Control of the arc plasma is best achieved by using a ⅛ in. (3.2 mm) diameter electrode having a flat tip from 0.060 to 0.100 in. (1.52 to 2.54 mm) diameter and tapered through a 10 deg draft angle.

Table 5—Length of Defects Repaired Horizontal Free-Fall Welding as a Percent of Length of All Weld Defects Repaired

Crew module	S/C 101	S/C 102	S/C CO 099
Inches of weld defect repaired—all welds	211	68	20
Inches of weld defect repaired—FF—Horizontal	208	58	8
Length of defects FF—horizontal repaired as percent of total weld defect length repaired	98.8	85.5	40.0

Table 6—Tool Type and Weld Position—Inches of Welding

Tool type— weld position	Approximate inches of weld		
	S/C 101	S/C 102	S/C CO 099
Hard-flat	720	615	615
FF-flat	402	402	402
Hard-vertical	647	647	593
FF-vertical	653	653	707
Hard-horizontal	318	423	423
FF-horizontal	1979	1979	1979
Total	4719	4719	4719

Filler metal (typically 0.060 in., i.e., 1.52 mm diameter) is purchased to specifications controlling chemistry and surface quality. It is cleaned and rewound (level wound) on 6 pound (2.7 kg) spools and packaged in plastic bags with dessicants. All filler metal is stored in a cabinet heated to 150°F (66°C) and humidity-controlled. When filler metal is placed on the welding head, it is mounted in a sealed, air-tight drum and driven by rollers through nylon tubes to the weld head. The outside layer of filler metal is run out and discarded as a further precaution against accidental contamination. After welding, the filler metal is removed, repackaged with dessicants, and stored in the same cabinet as described above.

Part and tool cleanliness are extremely important in avoiding porosity. Parts that arrive from machining sub-contractors have been cleaned, chemically filmed to within 3 in. (76.2 mm) of the edge to be welded, and coated with a strippable vinyl to give maximum corrosion protection during handling and storage. Prior to insertion of the parts in the tool, the tool surfaces and hold-down fingers are wiped with a cloth dampened with methyl ethyl ketone (MEK) followed by a dry cloth wipe to remove any residual surface films. The part is then stripped of vinyl and trimmed. The oxide is removed from the surfaces in the weld area by a rotating Bear-Tex* wheel. An MEK wipe, followed by a dry cloth wipe, are used to remove Bear-Tex* residue, and a vacuum is used to pick up the dust or oxide generated. Thereafter, surfaces are protected by aluminum foil, and white gloves are worn.

The final preweld clean consists of hand-scraping the four base metal surfaces 1 to 2 in. (25 to 50 mm) back from the weld joint. The faying surfaces are filed and a $1/32$ in. (0.79 mm) chamfer at approximately 45 degrees to the surface is made. A final vacuum is used. The contour bars are then adjusted to provide full contact with the part and joint gaps are closed. Verification is made that no joint gap exceeds 0.040 in. (1.02 mm) and are less than 1 in. (25.4 mm) long. Final inspection of part cleanliness is then accomplished first by bright light, then by black (ultraviolet) light. Once the weld process has begun, no penetrant inspection is permitted until the bead is finished and submitted to inspection.

To ensure maximum cleanliness, all welding operations are performed in an isolated, clean room area, originally used for final assembly and checkout of the Apollo. The room is controlled to temperatures of 74 to 78°F (23.3 to 25.6°C), and the relative humidty is controlled from 35 to 50%. If relative humidity gets too low,

*Trademark—Norton Coated Abrasives and Tape Division.

static electric charges will permit dust to adhere to parts; if humidity is too high, it may permit additional moisture to be adsorbed to surfaces. The clean room is a limited entry area with entry requirements for shoe cleaning and dust removal by blower. Though not required, it operates very closely to a class 340,000 clean room.

Quality Control Inspection

Welds are inspected by visual, radiographic, and penetrant techniques. X-ray provides for a 2% resolution of defects, but dye penetrant is far superior for uncovering surface defects. Weld beads are flushed prior to the penetrant inspection.

A P149 fluorescent penetrant (Uresco) system is used along with an E157 non-water washable emulsifier. This system provides for detection of defects to a level six sensitivity per MIL-I-25135. Experience has shown that crack defects that are 0.050 in. (1.27 mm) deep and 0.025 in. (0.64 mm) deep can be detected to a 90% probability, 95% confidence level using this system.

Weld Quality

The welding of the first crew module (Spacecraft 101) was completed during December, 1974. Welding of Spacecraft 102 was completed in September, 1976, and the third crew module (Spacecraft CO 099) was completed in February, 1980. A high overall level of weld quality has been consistently achieved. Defects have been essentially due to pores or stringers. Concave root surface (suckback) was a major problem on weld 35 on Spacecraft 101.

Table 2 tabulates the approximate inches of weld defect which required repair for each of the 36 welds. For the purpose of this paper, the dimensions of radiographic defects along the weld axis, not accepted by Material Review action, were totaled to determine "inches-of-weld defect repaired." Also, the inches of weld defect repaired does not include fluorescent dye penetrant defects as these were either of minor importance or were associated with radiographic defects which are included.

Study of Table 2 shows that in the main, welds 9, 28, 29, 34 and 35 offered welding problems for all three crew module structures. All of these welds were made using free-fall tooling. Table 5 graphically depicts that free-fall welding in the horizontal weld position was the source of most weld defects.

Tables 3 and 4 illustrate, by percentages, the overall high quality obtained in welding the three crew modules. The fact that the welding quality has been steadily improved is also illustrated by these tables.

Inertial Welding at Rocketdyne

by Sam L. Jones

A difficult welding job in a component of the engine for the space shuttle was made easier with the use of a modified inertial welding machine at the Rocketdyne division, here, of Rockwell International Corp.

The component is the main injector, and its construction involved the welding of 600 narrow posts into a circular concave cavity. The injector is the part through which the liquid oxygen is mixed with hydrogen en route to the main engine of the shuttle's booster.

Paul R. "Dick" Winans, manager of manufacturing product support in Rocketdyne's engineering department, said the use of an inertial welding machine is unusual in an aerospace application.

The machine had to be modified to provide more precise dimensional control for Rocketdyne's applications.

Rocketdyne adopted inertial welding as a production improvement over conventional welding, which was used initially. The new unit is several times faster than the old methods.

Developing an inertial welding system, at a cost of $425,000, also provided Rocketdyne with the ability to meet changes in the design or materials of the posts, Winans said.

Winans said, "If everything goes well, we can weld 600 posts in a couple of days, as against a week with the old machines. Based on our experience now, we cut out at least half the time. That was something we gained as a sideline. We had to get the capability to do the welding, and the side benefits are that we are doing it better and faster."

As the space shuttle program proceeds, there will be modifications to the main engines and the posts. Winans said, "We've got preliminary designs we're looking at for new posts".

Dissimilar metals, Inconel 718 and Hanes Stellite 188, are joined in welding the posts. Winans said, "We have to put in each post with high enough energy to make a successful weld with a minimum of upset and a minimum of flash".

Each post fits over a precisely designed stub out. The posts vary in length from about 8 to 10 inches, and each has a spline on it to take the torque of the welder's ram and a little shelf to take the ram's thrust.

The welder, Model 120VE, was modified to Rocketdyne specifications by the builder, Machine Technology, Inc., Mishawaka, Ind.

Winans said, "We try to buy stuff probably beyond the state of the art a little bit, as we did on this machine. We demanded that the machine builder manufacture things that he wasn't capable of doing at the time that he sold the machine to us, although he thought he could".

When the inertial welder was delivered, problems were encountered in the programming, such as integrating the X-Y table and the machine. Winans said, "And then with what we asked the machine to do in relation to the system the builder put into the machine, it didn't work well".

The problems were worked out over the next two months with the help from "a crew" from Machine Technology, he said.

Winans said, "We found with the capabilities and more sophisticated controls that we have, we do things a lot faster and a lot more reliable". Compared with the previous conventional welding equipment, Winans estimated that the newer system is about 30 percent faster and probably 50 percent more reliable, although precise data is not yet available.

Using the welder, a technician manually inserts each post for welding over the corresponding stub out. In the welder, computer controlled by a Giddings & Lewis computer and ASCI II language, a flywheel spins at 3,275 rpm. This is thrust against the hydraulic component, producing about 6,000 pounds of pressure (estimate at 30,000 to 40,000 psi). The computer locates the post beneath the welder's single ram based on the programmed design features.

Winans said the ram accelerates beyond 3,275 rpm, then comes down back through that speed, and trips the speed indicator on the way back down.

When the ram comes down, it makes the weld, then goes back up to the retract position. At that position, it activates the computer for the next position.

The computer moves the X-Y table so that the new post is positioned directly beneath. The speed decays in less than 3 seconds and no vibration occurs, Winans said.

INERTIAL WELDING machine used by Rocketdyne operates vertically. The two researchers at right are working with one of two computer consoles that control the welding and monitor the entire operation.

Blast off

Electron beam welding played an important role in the building of the weight-conscious Space Shuttle

By PAUL TURK
AMM Reporter

LOS ANGELES—In case you were wondering, the greater part of the National Aeronautics and Space Administration Rockwell International Space Shuttle system is built of aircraft aluminum, mechanically fastened with rivets and bolts.

Welding is much less in evidence. It is merely critical to two parts of the program—housing the crew and supplying the power to get the four-section assembly into—and out of—the air in the first place.

Fusion welding is used extensively to join sections of the large main propellant tank for the liquid hydrogen and oxygen used to fuel the shuttle orbiter's three main engines.

But the propellant tank is the only part of shuttle to be discarded during a mission. The two solid rocket boosters are recovered and reconditioned for use on subsequent missions. The orbiter itself is designed for 50 or more missions.

The orbiter is double-walled aluminum where the two-level crew compartment is concerned. The crew capsule or compartment is built separately as a welded aluminum structure, a pressure vessel, to enhance shirtsleeve environment for the crew in space.

The structure is fitted inside the fuselage, joining it at the sides, bottom, top and rear to the structure. The crew compartment is built with its own windshields, which butt against and match similar panels built into the overall structure. The windshield is a three-layer installation, for protection should an outer or inner windshield be damaged.

The most ambitious welding is in the powerhead of the Space Shuttle main engine, built by the Rocketdyne division of Rockwell International Corp., Canoga Park, Calif. The head is assembled with a deep, electron beam weld, for strength and to save weight, according to J.G. Somerville,

manager of material and process engineering at the division.

Rocketdyne's experience with welding in the engine program has convinced the division that welding will be the process of choice in many of its later programs, as well, Somerville said.

More than 200 major electron beam welds are made in construction of the 7,000-pound main engine. Some of the welds are an inch deep. Several other welding techniques are involved in construction of the main engine, and the original design, later modified, was for an all-welded construction.

There are about 400 inches of electron beam welding on the shuttle engine's powerhead, a three-dome structure to which the

rest of the engine's equipment is attached. Depths are from 0.020 to 1.25 inches in the Inconel 718 and Incoloy 903 heads, Somerville said. There are more than 30 inches of welding where the weld is an inch or more deep.

Electron beam welding, particularly deep welding, is used to minimize distortion of the parts, he said, and to eliminate the need for bolting the head as a backup to welding.

He said it was impossible to estimate the weight savings involved because "you design for it (welding)" and don't even consider bolting. He said the weight savings could be in the "hundreds" of pounds per engine, a significant savings in a weight-conscious vehicle.

A WELDING engineer at the Rocketdyne division of Rockwell International Corp. checks an electron beam gun and closed circuit television camera prior to computer-controlled electron bond welding of the Space Shuttle main engine combustion chamber. Automatic welding of parts with complex shapes is possible with this advanced welding capability.

189

Much of the Rocketdyne electron beam welding was carried out on a Sciaky 42KW, computer-controlled unit owned by NASA, but installed at Rocketdyne's Canoga Park facility for shuttle main engine production.

"We've done some MX work on it," Somerville said, but "the No. 1 priority is NASA, and right now, that's the Space Shuttle."

While the NASA Sciaky unit is the largest electron beam welder at Rocketdyne, there are four other, smaller Sciaky units and three Hamilton Standard machines used to support engine construction.

Engine fabrication and assembly, Somerville said, also involves gas tungsten arc, plasma arc, gas metal arc, inertia and resistance welding systems. "We need a range of processes," he said, to make the best use of the characteristics of each for different elements of the complex, powerful engines.

Inertia welding, where two parts are spun against one another until they fuse, is a major technique in use on the main engine. The main injectors are attached in that fashion, and some vibration problems have been encountered in that area, Somerville said.

Inadequate baffling of the injectors, which introduce preburned fuel to the chamber in which the oxygen and hydrogen are mixed to produce thrust, was producing a vibration which was damaging the parts.

The injectors were redesigned using additional baffling, and the problems appear to have stopped. But the injectors, long narrow tubes, are subject to great stress, and since the shuttle and its engines are expected to last 50 missions or more, there are refurbishment requirements, and that could include removal of damaged or suspect injectors.

Even with the inertia weld, Somerville said, Rocketdyne would remove and replace the injectors by machining off the welds, re-preparing the surface and re-welding them into place.

From a design standpoint, Somerville said, much more welding is involved on the Space Shuttle engine program than on the earlier engines Rocketdyne built for other programs, including the Saturn engines used to launch the Apollo series of moon missions.

The welding emphasis probably will continue in future engines, he said, because of reduced weight and the appropriate reliability of the techniques used in the engine. There have been no particular problems with the various weldments in the engine during its tests, even as other difficulties have hampered its development and testing until recently.

The Space Shuttle engine, Somerville said, started as a totally welded structure, in fact, but some changes to mechanically-fastened segments were made for access and for other reasons not necessarily connected to concern about the adequacy of welding processes and their ability to withstand the engine's strains.

Welding for spacecraft engines, Somerville said, "came into full flower on the Space Shuttle's engines."

Presented at the CASA/SME Autofact West Conference, November 1980

NC Machining of Wind Tunnels

by Dr. N. Akgerman
C.F. Billhardt
Battelle-Columbus Laboratories
and
D.G. Bowen
NASA - Langley Research Center

ABSTRACT

NASA-Langley is called upon to make wind tunnel models of advanced air-craft designs. Their problem was that their APT IV system could not de-termine the intersection curve between the wing and the fuselage of a model. In addition, the data defining the wing and fuselage was generated by computer analysis programs and the data did not lend itself to direct input to APT. Battelle developed a Fortran program which would accept the input data in more than six different formats, generate the cutter locations for the entire structure including the wind/fuselage intersection, and out-put a CLTAPE ready for conventional post-processing.

Since being installed, NASA has also used the program for developing NC tapes for strut/nacelle and canopy/fuselage structures. Battelle has also used the program for machining models of steam and gas turbine blades.

INTRODUCTION

One of the functions of NASA-Langley Research Center, Hampton, Virginia, is to develop and test advanced designs for aircraft and other flight structures (such as the space shuttle). The geometry of these structures is derived from computer models which can make preliminary performance analyses of structures given some set of flight conditions (speed, range, altitude, etc). Designs which appear promising are then further evaluated by wind tunnel testing. Converting the numerical data from the computer models into a physical model is one of the primary responsibilities of the Langley Machine Shop.

This group had APT available to them on a CDC multi-system computer installation. This was used, where possible, to generate tapes for NC machining. There were two major problems with using APT, however. The first is that the wing and fuselage of a model were essentially defined as two separate items. That is, a fuselage would be defined as though it were simply a cigar-shaped object with no thought given to the wing projecting from it. Similarly, a wing would be defined from the centerline to the tip, with no thought to the fuselage to which it would be attached. Thus, the curve of intersection of the two elements was undefined. Figure 1 illustrates the complex nature of a typical wing-fuselage intersection curve.

This lack of knowledge of the curve of intersection required the Langley staff to manually determine the approximate intersection area. They would use APT to generate NC tapes for both the wing and fuselage up to this area. The unmachined intersection area would be finished by manual machining and hand benching using an experienced machinist's judgment as to where the intersection was.

The second major problem area was the format of the data from the analysis programs. This was generally in tabular form and had to be pre-processed in order to be handled by APT.

As a result of these two problem areas, and the demands to provide more accurately machined models for wind tunnel testing, in early 1978 Langley contracted with Battelle to develop a computer program which would define the CL (cutter location) points for a combined wing and fuselage structure, including the intersection of the two components. The result was the program AIRCL. As input, this operates on data files generated in 7 different formats by flight simulation programs. The output is a file which emulates the format of CDC-APT Section 2. Langley passes this file through their APT compiler along with the specifications for the NC machine to be used in order to obtain the post-processed file, ready for tape punching and execution.

AIRCL is divided into the following major processing stages:

o Read the input data according to format specifications, and perform logical tests on array sizes and position of data.

o Perform preliminary data interpolation and calculate cutter offsets for the wing and fuselage as separate items.

o Find the curve of intersection between the wing and fuselage.

o Generate the output CL points, limiting the fuselage and wing by the intersection between the two and interpolating to achieve the specified surface accuracy.

o Generate a trim cut, if specified, which will machine around the plan view outline of the model.

FINDING CUTTER OFFSETS

The direction of a cutter offset point is found by forming the cross-product of two vectors tangent to the surface. In the case of a wing surface, the points used to form the tangent are the points on either side of the one in question, along the X-axis. Because of the extremely swept-back nature of some wings, this technique would not work for the Y-axis tangent. In this case, the wing station on either side of the point is interpolated to find the Y and Z values at the same X as the point in question. These three components are then used to find the second tangent and the cross-product of the two normals yielding the direction of the normal.

The magnitude of the cutter offset is the specified cutter radius to which an additional offset is added to prevent gouging or undercutting. This additional offset is a function of local geometry and is calculated uniquely for each point.

FINDING THE WING/FUSELAGE INTERSECTION

The curve of intersection between the wing and the fuselage is found by an interactive process. Referring to Figure 2, the process is as follows for the I-th wing pass (WP_i). A fuselage bulkhead, defined by Y-Z coordinates, is found by interpolation at the X coordinate where the wing pass intersects the fuselage centerline. The point of intersection (Y_p, Z_p) of the wing pass and the trial bulkhead is then located. If the Y coordinate where the bulkhead intersects the wing pass in the Y-X plane is the same (within a small tolerance) as the Y coordinate of the inter-section of the bulkhead and the wing pass in the Z-Y plane, the point of intersection has been found. This is unlikely to occur on the first trial. A second trial bulkhead is then established at the X coordinate on the wing pass corresponding to the trial point of intersection. The process con-tinues looping in this manner until the Y value of the point of intersection in the Y-X plane corresponds to the Y value in the Z-Y plane.

It should be noted that the approach taken in AIRCL is:

1. Determine the CL-surface for the wing and fuselage separately.

2. Determine the intersection of the two CL-surfaces.

3. Eliminate CL-points inside the curve of intersection.

This contrasts with the traditional approach which is:

1. Determine the intersection of the wing and fuselage surfaces.

2. Calculate the cutter paths, taking great care not to gouge the wing when machining the fuselage and visa versa.

The AIRCL approach eliminates the complexity of the second step in the traditional method. Finding the intersection of the CL surfaces, rather than the part surfaces, ensures that the cutter is tangent to both part surfaces and effectively eliminates the gouging problem.

Special circumstances such as the wing pass being ahead, behind, above or below the fuselage are also tested for the appropriate action taken.

SURFACE FINISH

One of the options available to the user is to specify the scallop height between adjacent cutter passes. Knowing this and the cutter size, the distance between passes is found. The length of each fuselage bulkhead and each wing section is calculated to find the longest of each. These maximum values are divided by the distance between each pass required to meet the specified surface finish. This yields the total number of passes required on both the wing and the fuselage. Each fuselage bulkhead and each wing section is then divided into the appropriate number of pieces. Table 1 shows the number of passes required per inch with different size cutters to achieve various uncut scallop heights.

The internal storage arrays in AIRCL are sized for a maximum of 150 passes on the wing and 50 passes on the fuselage. If the number of passes on either part is greater than the respective maximum, a direct access data file is established. This permits, on the wing for example, each chord to be interpolated, one section at a time and the resulting points written to the file. When all sections have been interpolated, the cutter paths are formed by picking the I-th point from each section. These points then define a cutter path from the intersection curve to the wing tip. AIRCL is capable of generating sufficient passes so that a wing chord of 8 feet is machined with a residual scallop height at any point of 0.001 inch.

PROGRAM STRUCTURE

The AIRCL program is written entirely in FORTRAN with the exception of three subroutines which handle the direct access storage file. The program was developed and tested on Battelle's CDC-6500 computer, then transferred and implemented on Langley's multi-CDC system. The program modules were linked using CDC's Segmentation Loader. This is a form of overlaying, intended to reduce the memory size required for program execution. The program, when linked, requires approximately 41.5K (decimal) words (60 bit word length) of central memory.

DATA INTERPOLATION

One of the primary requirements given by Langley during development of AIRCL was that the cutter paths accurately reproduce the surface defined by the input data. Where interpolation was used, the resulting surface must be smooth and pass through the defined points. The program was coded to allow the user to specify the type of interpolation to be used, with different types of interpolations used for different areas. The areas or directions, for which interpolation can be defined are:

(1) On the wing along the chords

(2) On the wing from centerline to tip

(3) On the fuselage along the bulkheads

(4) On the fuselage from nose to tail

(5) Along the fuselage-wing intersection curve.

In addition, for the centerline to tip wing passes, different interpolations may be specified between different wing sections. This permits
a smooth spline interpolation to be used on different sections of a wing
with a sharp break in it, with the actual transition handled by linear
interpolation. A prime example of the need for this capability is the
gull-wing of the World War II F4 Corsair fighter.

The types of interpolation possible are linear, spline, and multi-
order polynomial. The default is spline for all except the intersection
curve which uses linear as default.

OPERATION OF AIRCL BY LANGLEY

As noted, AIRCL generates an output file which emulates the output of
APT, Section 2. When originally implemented in 1978, Langley would pass
this file through APT Section 3 and 4 on their CDC computer, then dump the
output to punched tape. As is the case in many large computer installations, the machine shop was the only group which used punched tape. The
tape punch was old and difficult to service and the computer center wanted
to eliminate it. In 1979, the machine shop installed a Uniapt NC
programming system. Since then, the output of AIRCL is copied to a magnetic tape. This is then transferred to the Uniapt system for post processing and output to punch tape. Figures 3 and 4 illustrate models made

at Langley using AIRCL.

OTHER APPLICATIONS OF AIRCL

Although originally designed specifically for wind tunnel models defined by wing and fuselage data, AIRCL can handle any two surfaces defined in a similar manner. By rotating the data, NASA has used it to machine aircraft tail control surfaces, and engine nacelle and pylons. Battelle has used it to machine turbine blade models. In this case, because of the complex, non-smooth nature of the blade roots, a number of enhancements were added to the original program. Figure 5 is a turbine blade model made using AIRCL.

PROGRAM AVAILABILITY

The original source code for AIRCL is available to the public. Information regarding how to acquire the program may be obtained by writing:

Computer Software Management and Information Center

112 Barrow Hall

University of Georia

Athens, Georgia 30602

Inquiries should reference computer program LAR-12494.

SUMMARY

AIRCL is a general purpose program which develops the CL file for NC machining wind tunnel models or similar items. With the fuselage and wing defined as separate entities, AIRCL determines the intersection of the two, then generates the cutter coordinates for each surface which lies outside this intersection. The program has been in production use for two years. Although some minor bugs were discovered after installation, the program was so structured and documented that Langley was able to correct these themselves.

AIRCL allows Langley to develop punched tapes for complete aircraft models with no direct NC programming effort. The NC tapes are so complete and accurate that in many cases, only hand polishing of the surface is required before the model is ready to be flown.

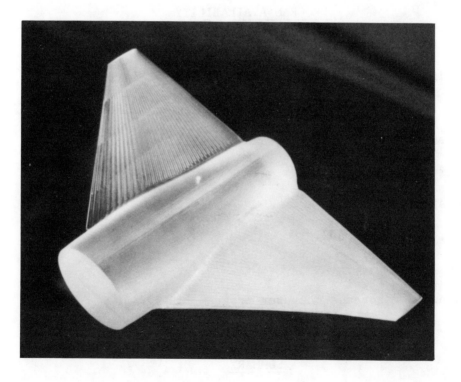

Figure 1. Typical Wing-Fuselage Intersection Curve

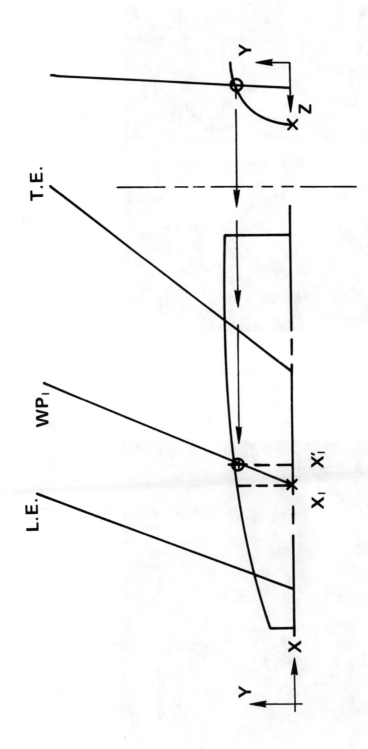

FIGURE 2.

FINDING THE WING-FUSELAGE INTERSECTION

Figure 3. Wind Tunnel Models Made Using AIRCL

Figure 4. Detail of Typical AIRCL Model

Figure 5. Turbine Blade Model Made Using AIRCL

TABLE 1.
PASSES PER INCH REQUIRED TO PRODUCE SURFACE
WITH SCALLOP HEIGHTS AS SPECIFIED

SCALLOP HEIGHT (IN.)	CUTTER DIAMETER (INCH)			
	0.250	0.500	0.750	1.000
0.040	5.46	3.68	2.97	2.55
0.020	7.37	5.10	4.14	3.57
0.010	10.21	7.14	5.81	5.03
0.005	14.27	10.05	8.19	7.09
0.001	31.69	22.38	18.27	15.82

A Computerized Ultrasonic Multiple Array Real-Time Inspection System for the Space Shuttle External Tank Liquid Hydrogen Barrel Welds

by Richard G. Harrington
Martin Marietta Aerospace

ABSTRACT

The Space Shuttle External Fuel Tank contains over 3,000 linear feet of fusion-welded aluminum, which is evaluated by conventional nondestructive testing methods. The present weld evaluation methods consist of both radiographic and liquid penetrant inspections, which are cost-intensive due to time, labor and materials involved.

In order to overcome this problem, Martin Marietta Aerospace Michoud Division developed, under a NASA Technical Directive, a prototype computerized ultrasonic multiple array weld inspection system with potential application to the External Tank.

The prototype ultrasonic weld inspection system will be evaluated in a production environment on the liquid hydrogen barrel welds, and compared to the present x-ray and penetrant inspection methods for six to eight months.

This paper will outline the unique design, equipment involved and the operational aspects of the computerized ultrasonic weld inspection system for nondestructive evaluation of aluminum welds. The system includes an ultrasonic probe array consisting of twenty ultrasonic transducers, a MatEval MicroPulse Unit that can handle up to 240 transducers, an HP-9845B computer, a weld mismatch and peaking measuring device, and a scanning head.

INTRODUCTION

The Space Shuttle External Tank is fabricated by the Martin

Marietta Corporation at the NASA Michoud Assembly Facility located 15 miles east of downtown New Orleans, La.

This giant fuel tank is composed essentially of two tanks; a large liquid hydrogen tank and a smaller liquid oxygen tank, connected by an intertank skirt. This forms one large storage container (154 feet long and 27.5 feet in diameter).

The liquid oxygen tank, which is the External Tank's most forward section, is joined together by 1,052 linear feet of weld. The liquid hydrogen tank is the External Tank's largest component and consists of 1,995 linear feet of weld. The weld footage for the oxygen and hydrogen tank is greater than one-half mile, or more specifically, 3,047 linear feet.

The External Tank has two major roles in the space shuttle program:

1. To contain and deliver liquid hydrogen, and liquid oxygen to the orbiter engines.
2. To serve as the structural backbone of the space shuttle during launch operations and the early stages of flight.

PRESENT WELD INSPECTION METHODS

Every inch of the External Tank weld is presently nondestructively tested, using both x-ray and penetrant inspection methods. These methods together require a minimum of 6 man-hours to inspect 20 feet of weld on the liquid hydrogen barrel. This includes the following:

- Penetrant Inspection
- X-ray film layup and setup of equipment
- Operation of x-ray equipment
- Removal of x-ray film
- Processing of film
- Interpretation of x-ray film

In addition to x-ray and liquid penetrant inspection, all welds are inspected for mismatch and peaking. Mismatch is the offset of the base metal at the weld center line. Peaking is either positive or negative angular deviation of the base metal at the weld center-line.

Mismatch and peaking measurements are presently performed by two methods:

1. A manual method using a comb, protractor, and optical comparator.

2. A hand-held, computerized, mismatch and peaking measurement
    system developed by Martin Marietta's Advanced Quality
    Technology Dept.

All the present nondestructive testing methods for inspecting
the External Tank welds, including mismatch and peaking are cost-
intensive, due to time, labor and materials involved.

FEASIBILITY STUDY/AUTOMATED NDE

To help overcome this problem, Martin Marietta's Advanced
Quality Technology Department performed a feasibility study in 1976
to evaluate Automated NDE techniques. As a result of this study,
Martin Marietta developed a small, 4 channel, computerized ultrasonic
weld inspection system, which proved feasible for possible application on the Space Shuttle External Tank Welds.

PROTOTYPE WELD INSPECTION SYSTEM

The development of the 4 channel computerized ultrasonic weld
inspection system led to the fabrication of a 20 channel prototype
system which is to be evaluated on the liquid hydrogen barrel welds.
The system was built to Martin Marietta specifications by MatEval
Ltd. in Warrington, England.

The prototype system consists of:

- A 20 channel ultrasonic probe array
- A MatEval MicroPulse unit
- A Hewlett Packard 9845B computer
- 4 Linear Variable Differential Transformers (LVDT)
- A flaw marker pen
- A scanner head and support arm
- A drive system and stepper motor
- A verification test panel
- A transport cart, water tank and suction pump

ULTRASONIC PROBE ARRAY

The ultrasonic probe array contains 2 longitudinal wave transducers and 18 shear wave transducers. The 2 longitudinal wave
transducers are used to monitor water coupling and measure weld land
thickness.

The 18 shear wave transducers are used for detection of
surface and internal defects, and have been optimized at 45 degree
and 60 degree angles in the probe block for the liquid hydrogen

barrel weld geometry. Each transducer is multiplexed (sequentially switched) in a predetermined sequence, providing a series of pulse echo and pitch and catch scans.

Demineralized water is used as a couplant for the ultrasonic transducers and transmits the ultrasonic energy from the probe array to the part being tested. The demineralized water is re-circulated by a suction pump from a small reservoir located on a transport cart to the polystyrene probe blocks where it is channeled under the probe array. Two rows of bristle brushes surround the probe array, contain the demineralized water under the probe blocks, and remove excess water coupling during inspection. This method works very well with no loss of coupling or water leakage, even while inspecting at a 45 degree angle.

MATEVAL MICROPULSE UNIT

The MicroPulse unit is composed of two parts:

1. The "local site" which is located on the transport cart and contains modules which control the inspection operation.
2. A "remote site" which is attached to the automated weld inspection system drive assembly and linked to the "local site" by an umbilical cable. This reduces or minimizes signal degradation from the ultrasonic probes.

A microprocessor in the micropulse "local site" controls the inspection sequence by multiplexing the transducers collecting the test results, and passing the data to the HP-9845B computer for analysis and storage on floppy disk. Similarly, the micropulse unit collects data from the peaking and mismatch measuring LVDT'S and transfers this to the Hewlett Packard computer.

The micropulse unit can separate parameters for up to 256 different tests and can be hooked up to as many as 240 ultrasonic transducers. The microprocessor in the micropulse unit connects each transducer in turn to a separate flaw detector. Each flaw detector can have a different timebase, sensitivity, gate position and other normal ultrasonic parameters. Therefore, each test is individual, and can be set to perform different inspection operations.

HEWLETT PACKARD 9845B COMPUTER

The Hewlett Packard computer is used to control the entire weld inspection system. At the start of an inspection the operator inputs standard identification data via the keyboard. The equipment is calibrated and verified by scanning the verification test plate. During inspection, thickness monitoring and mismatch and peaking reports are generated in real-time, and all out-of-tolerance locations are indicated by the flaw marker.

On completion of an inspection a "data base" of indications is created from which tabular and graphical outputs are produced. The tabular format shows the axial position of the defect, its beam path position and echo amplitude. The graphical output consists of a 1:1, continuous scroll, B & C scan. In addition, a 4 x magnification and/or a transverse B scan can be produced for any desired weld slice. The results are displayed on both the CRT and computer printer.

LINEAR VARIABLE DIFFERENTIAL TRANSFORMERS

The 4 linear variable differential transformers (LVDT'S) are mounted across the trailing end of the scanner head. These are arranged in a line at right angles to the weld, so that there are 2 LVDT'S on each side of the weld. Each pair of LVDT'S has the same distance between their center lines.

From the reading of each pair of LVDT'S, a plane coincident with the welded surface is calculated. The angle of intersection of these planes is printed out as the peak angle.

$$PEAK = \emptyset = ARCTAN\ B-A/K + ARCTAN\ D-C/K$$

The linear dimension of this intersection point, above the nominal weld center line is printed out as the mismatch.

$$MISMATCH = A-C-X/K\ \ (B-A-D+C)$$

FLAW MARKER PEN

The flaw marker pen is mounted at the trailing end of the scanner head near the LVDT'S. The pen is solenoid-operated and will mark a water-washable ink on the panel adjacent to the weld land area when defects are detected by the micropulse unit.

SCANNER HEAD AND SUPPORT ARM

The scanner head has a supporting arm which terminates a flange and connects to the traversing drive system. The supporting arm carries the scanner head and can be pulled clear of the inspection area. The scanner head straddles the weld and has 4 wheels which run along the weld land to guide the probe array along the center of the weld.

DRIVE SYSTEM AND STEPPER MOTOR

The present in-motion x-ray fixture, for inspecting the liquid hydrogen barrel welds, uses a drive assembly with a D. C. motor. A new drive assembly was built with both a D. C. motor and a stepper drive motor to run the ultrasonic system. The stepper drive motor allows the ultrasonic system to step at 0.040 inch increments and inspect at 12 inches per minute over nondefect areas.

VERIFICATION TEST PANEL

Prior to every inspection, the system is calibrated to a 5/64th through-drilled hole, and all 20 ultrasonic transducers and their corresponding tests are checked. Once the inspection is completed the verification test is performed again to assure that all transducers operated during the inspection. The verification test panel is mounted on one end of the actual liquid hydrogen barrel panel and a run-off panel is mounted at the opposite end to allow inspection of the entire weld.

TRANSPORT CART

A hand-pulled mobile cart is used to transport the Hewlett Packard computer, disk drives, micropulse "local site" unit, water tank and suction pump. This allows the cart to be the control center for the entire inspection operation.

The prototype system incorporates all the major characteristics of the present weld inspection methods, together with additional features, to provide a fully automated system. This means that only a single inspection will be necessary for any one weld, since thickness monitoring, defect detection and mismatch and peaking measurements are performed in one run.

LABORATORY EVALUATION

During the development of the Automated Weld Inspection System, a 7' x 3' test panel and test fixture was fabricated to evaluate the system. The test panel contained a series of through-drilled holes ranging from 0.020 to 0.100 inch in diameter and two 0.047 inch diameter through-drilled holes spaced 0.240 inch apart (edge to edge) to check system sensitivity and resolution.

In addition to the through-drilled holes, 24 electro-discharged machined (EDM) slots 0.050" long by 0.025" deep by 0.004" wide were installed in the weld crown, root and parent metal to check defect detection and coverage. The EDM slots were oriented:

- parallel to the weld axis
- perpendicular to the weld axis
- + and - 45 degrees to the weld axis

The test panel was mounted in the test fixture and the system operated. The system exceeded requirements, detecting all of the through-drilled holes and EDM slots. In addition, the system resolved the two 0.047 inch diameter through-drilled holes. The flaw marking pen located in the scanner head marked the actual locations of all the defects in the test panel within 0.125 inch.

PRODUCTION EVALUATION

The prototype system is now scheduled to be installed on the liquid hydrogen barrel weld fixture to evaluate the system in a production environment. The system will use the existing in-motion x-ray drive fixture to allow either in-motion x-ray or Automated Ultrasonic Weld Inspection on the liquid hydrogen barrel welds. This will allow the present in-motion x-ray inspection to continue, while performing the evaluation and confidence level testing on the automated ultrasonic weld inspection system.

CONCLUSION

The prototype automated weld inspection system is expected to provide benefits such as:

- Reduced manufacturing process time
- Improved weld repair quality
- Reduced x-ray film cost, processing and interpretation
- Significant cost savings in nondestructive evaluation for the External Tank welds.

CHAPTER 4

TOOLING

Reprinted from *Welding Journal*, February 1977

BIG TOOLS FOR DEEP SPACE

include the 125,000 lb "cigarette holder" fixture which measures 45x31x48 ft as well as a LH$_2$ tank welding fixture weighing 329,000 lb and a LO$_2$ tank welding fixture which weighs over 300,000 lb

An artist's view of how the Space Shuttle will jettison its two solid fuel rocket boosters 27 miles above the earth's surface. The Shuttle will afterwards continue into space with its External Tank and Orbiter, looking somewhat like the Graf Zepplin dirigibles of an earlier era

BY ROBERT M. POWERS

Moving man further into the environment of space, especially in any significant numbers, requires that the economics of space travel be a foremost consideration. Apollo, for example, delivered a pound (0.45 kg) of payload into orbit for around $1,000, a figure far too high for the future when 65,000 pound (29,484 kg) payloads will be common.

ROBERT M. POWERS is a free-lance writer from Wheat Ridge, Colorado.

With the end of Apollo, NASA planned for the next major step in space to be the Space Shuttle. The primary objective of the Space Shuttle is to provide a vehicle that will take off like a rocket, cruise and perform in space like a spacecraft, and land on a conventional runway like an airplane.

The Space Shuttle is designed to reduce the cost of space operations substantially and provide the ability to support many scientific, commercial, and defense activities. The Shuttle will lift one pound (0.45 kg) into orbit for only $100. The substantial reduction in cost afforded by the Shuttle is because most of the major components will be recovered for reuse rather than expended as was the case with past space exploration.

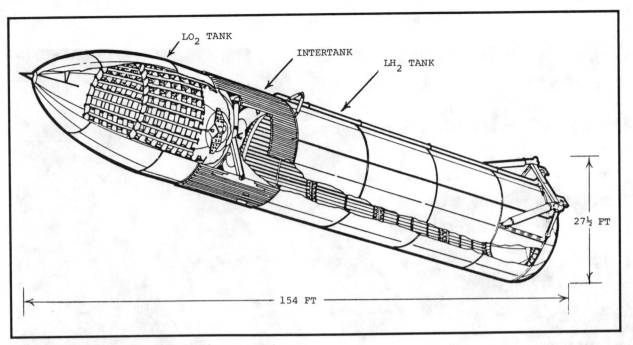

LO₂ TANK
INTERTANK
LH₂ TANK
27½ FT
154 FT

Fig. 1—Schematic of the External Tank with some statistics

The Space Shuttle consists of three major components:

1. The Orbiter for the crew and payload.
2. Two solid rocket boosters which assist in lifting the vehicle to an altitude of 27 miles.
3. The External Tank which provides the fuel for the Orbiter.

Not only must the External Tank carry the fuel for the Orbiter, but it is the main structural element of the vehicle at launch, supporting the solid rockets and the Orbiter.

The solid rocket boosters are jettisoned at 27 miles (lead photograph) and are parachuted into the ocean to be recovered, refilled, and reused. The External Tank goes into Earth orbit with the Orbiter and is then jettisoned. It also drops into the ocean at a programmable place, but it is not recovered because it disintegrates from heat as it tumbles. The Orbiter completes its assigned mission and reenters the atmosphere to land on a runway and be reused for successive flights.

External Tank Construction

Martin Marietta Corporation is building the External Tank at the NASA Michoud Assembly Facility in New Orleans, Louisiana. NASA's Marshall Space Flight Center at Huntsville, Alabama controls the project. The first tanks will be delivered in early 1977. After ground tests and flight qualifications, NASA expects to place orders for production of 45–60 tanks each year in the 1980's with the total number of Shuttle flights to the century's end nearing 600-700.

The External Tank, one required for each launch, is the first item of space hardware to be designed on a large production basis.

Some Numbers

The External Tank has a 27.5 ft (8.4 m) diameter and is 154 ft (46.9 m) long—Fig. 1. It weighs more than 75,000 pounds (34,019 kg) with a volume of 73,000 cu. ft (2,067

m³). In flight, the External Tank will hold 1.5 million pounds (653,886 kg) of liquid oxygen (LO₂) and liquid hydrogen (LH₂) propellant, feeding the mixture to the engines of the Orbiter until the craft reaches space.

Each tank is made of 2219 aluminum (temper T-87) with the primary thicknesses varying between 0.140 and 0.550 in. (3.6 and 14 mm). The welding (3,047 ft or 929 m done in three passes for a total of over 9000 ft or over 24430 m of welding operations) is accomplished using the gas tungsten-arc process with dcsp power and helium shielding. The filler metal for welding is 2319 aluminum alloy.

Fig. 2—Major components of the "cigarette roller" welding fixture

To manufacture this gigantic tank, basically composed of an LH₂ tank, intertank, and an LO₂ tank with an ogive nose, it was necessary to design and fabricate or use 364 special tools, of which, 34 are major fixtures. The largest of the tools are the LH₂ and LO₂ tank weld fixtures—essentially giant lathes for turning the tanks while welding them into a complete, integrated, and concentric whole.

Beginning Tank Assembly Procedures

Assembly begins when curved "pie-slice" gore segments are welded on a special fixture called a "gore-to-gore" trim and weld tool. Three gores form a quarter dome, two quarters are mated to form a half, and the halves are welded on another fixture to complete a dome. Since the External Tank is really two tanks, separate and complete, the ends of the tanks are domes.

The "Cigarette Roller"

The barrel sections of the External Tank are rolled and welded in a giant machine nicknamed the "cigarette roller"—Fig. 2. A section of aluminum plate is inserted into the machine and, one-by-one, other plates are welded to it and rolled around a cylinder until, as shown in Fig. 3, a completely welded and integral barrel is made—27.5 ft (8.4 m) in diameter.

Each barrel "rolled" on the machine is 21 ft (6.4 m) long. Three of these rolled barrels (each barrel has 8 individual panels) are used in each liquid hydrogen tank. Forty-eight edges are machined and 504 ft (154 m) of welding is done to make up these barrels. The "cigarette roller", one of the three largest of the fixtures (tools) which are used in the making of the External Tank,

Fig. 3—A 21 ft long barrel after it was welded with the "cigarette holder" fixture

weighs about 125,000 pounds and is 45 ft (13.7 m) wide, 31 ft (9.4 m) high, and 48 ft (14.6 m) long.

"Ogive Tools"

Although the liquid hydrogen tank (LH₂ tank) is basically a simple cylinder—though large, the forward tank, the liquid oxygen (LO₂) tank, is a cylinder with an aerodynamic shape. Since the LO₂ tank is the part which will have to strike into the atmosphere first, it has a nose piece called an "ogive" (ogive refers to the bullet shape of the nose piece).

To make the complicated nose for the LO₂ tank, Martin Marietta uses two large fixtures called "forward ogive" and "aft ogive" tools. These complicated tools weld and trim the oddly shaped panels of aluminum into the nose shape. The "ogive" nose of the External Tank represents the forward 32 ft (9.8 m) of the LO₂ tank. The forward ogive tool handles eight individual aluminum panels, each 15 ft (4.6 m) long. The aft ogive tool takes twelve panels, each 17 ft (5.2 m) long. Using these two tools, the front portion of the External Tank is trimmed and welded.

Final Tank Assembly

Precise and large, complicated and technical as the beginning tank assembly tools are—the dome tools, the ogive fore and aft tools, and the giant cigarette roller—there are two tools used for the production of the External Tank which dwarf the floor, the building, even the tank for which they were constructed. These are the tank weld fixtures which are used to weld together all sections of the tanks. Like a lathe, they take one of the tanks (there is a tool for the LH₂ tank and one for the LO₂ tank) and put it between chuck and tailstock and turn the tank as it is being welded.

Fig. 4—A LH₂ tank during process of being assembled on the LH₂ tank welding fixture

The tank welding fixtures were installed with a foundation sunk 35 ft (10.7 m) into the ground beneath the Michoud Assembly Facility's factory floor. The LH_2 tank major welding fixture weighs 329,000 pounds (149,232 kg) above the machine ways—Fig. 4. It is 30.5 ft (9.3 m) high, 166 ft (50.6 m), long and 55 ft (16.8 m) wide. The LO_2 tank major weld fixture is 125 ft (38.1 m) long, 30.5 ft (9.3 m) high, and has a width of 55 ft (16.8 m); it weighs over 300,000 pounds.

On these two major weld fixtures, the headstock alone weighs 138,000 pounds (136,078 kg) and the tail stock 53,000 pounds (24,040 kg). The two tank weld fixtures are designed to align, precision trim, and join by welding the individual major tank assemblies including domes, barrel sections, intermediate frames, and a nose section for the LO_2 tank.

Despite the massive size of the weld fixtures, they operate on a pair of precision ground ways which enable the tool, with its other features operating, to hold dimensions to a tolerance of 0.010 in. (0.25 mm) over a range of 30 ft (9.1 m). Accuracy is assured by the use of laser/optical measurement techniques combined with precision jacking and other features of the tools.

Unique to the lathe-like nature of these two tools is an expanding mandrel which acts as a sizing media and weld heat sink at each of the welds. The expanding mandrel is 27.5 ft (8.4 m) in diameter, and can expand the aluminum alloy (Alloy 2219, T-87 temper) barrels as much as 3/8 in. (9.5 mm). The savings in time over a manually positioned segmented mandrel such as the closeout mandrel is 360 manhours per tank.

In addition to the dual shaft-mounted, remote deployment mandrels of the weld fixtures, they also have another unique device: there is a combination weld gantry/X-ray shield, the tallest and widest structure on the fixture. The gantry completely straddles the basic fixture elements and the tank being welded to provide a stable platform for the weld equipment and torches. The gantry also serves as a radiation shield when in-process X-rays are taken to confirm weld quality.

Welding of the tank sections is accomplished by rotating the tank past a fixed position set of welding torch heads. The torch heads are located in a nearly downhand position which leaves the weld torches and weld operators 27 ft (8.2 m) above floor level. After a typical set of circumferential welds are completed (three passes are made on each weld—an initial sealer pass, followed by a continuous penetration pass and a final filler pass) and the gantry is still in position, X-ray film is wrapped around the full periphery of the tank weld and an X-ray source inside the tank is activated. The gantry is fully lined with lead to serve as a radiation barrier. This "in-process" nondestructive evaluation of the welds does not impact on related areas of production flow. The gantry rides on its own pair of railroad rails which are located 45 ft (13.7 m) apart and straddles the main fixture precision ways. It is capable of moving up and down the length of the fixture and tank to any required position for welding together tank elements.

Welding Process Characteristics

The weld equipment for the fixtures was fabricated by the Sciaky Corporation of Chicago, Illinois. Solid-state electronic power supplies are utilized to provide constant current. The voltage is controlled to ± 0.1 volt by using voltage-controlled welding heads.

The welding speed and filler metal feed speed are controlled by solid-state electronics. A computerized digital taper control commands the solid-state welding system, programming the four welding variables—amperage, voltage, travel speed, and filler metal feed speed—as a function of torch travel and is being used for the welding of tapered thicknesses of the tanks. This control also allows for memory and documentation, both of the welding schedule, which may be either preset or manually developed during performance of the weld, and of the actual output of the welding system as compared with programmed commands.

The width of the welding passes, especially the filler pass, is controlled by a magnetic probe whose field can be varied in frequency and magnitude. The top and under sides of the weld beads are very smooth and uniform so that no shaving of the beads is necessary.

Post Assembly Activity

When the giant tools have done their jobs, the tanks are tested, sprayed with an ablative foam (a material which will protect the surface of the External Tank from atmosphere friction) and then checked out. When the LO_2 tank, the connecting intertank, and the LH_2 tank have been finally assembled, the 154-foot long External Tank is loaded on a barge and shipped out the intercoastal waterway from New Orleans to the Gulf of Mexico. From the Gulf, the NASA barges, *Orion* or *Poseidon*, with their cargo of one very large External Tank, go around the tip of Florida to enter the Indian River from the Atlantic Ocean and unload the Tank at Cape Canaveral and the Kennedy Space Center.

At the Center, the External Tank is mated to a waiting Orbiter and two solid rocket boosters. From there, it is a small step into space for up to six "passengers" and the four crew members of the Space Shuttle.

Acknowledgement

All illustrations were supplied through the courtesy of Martin Marietta Aerospace, Michoud Operations, New Orleans, Louisiana.

Simple tooling positions tube welds

Arc-welding the multitude of tube joints in the cramped quarters
of the Space Shuttle orbiter called for simple but innovative tooling

THE TUBING in the forward reaction-control system, hydraulic system, and auxiliary power units of NASA's Space Shuttle orbiter is joined with some 550 gas tungsten arc welds, the soundness of which is critical to the spaceship's control and operation. The reaction-control system provides the thrust for attitude maneuvers and for small velocity changes when the orbiter is above 70,000 ft. The hydraulic system provides pressure to various equipment, including the three main-engine gimbals, main-engine valves, landing gear, and brakes. The auxiliary power units provide mechanical power to drive a hydraulic pump.

Although some of the tube welds are made in benchtop setups, about 75% have to be made inside the orbiter fuselage after much of the mechanical, electrical, and electronic components have been installed. Welders must not only gain access to the tube joints but also avoid damage to the sensitive components during welding.

As shown in Fig 1, all of the circumferential welds are made through a sleeve surrounding the butted tubes. 100% fusion is required, and no suckback of the weld on the ID of the tubes is permitted. The presence of a slight weld bead on the ID, which can be detected by radiographic inspection, is preferred as assurance of full weld penetration. Each joint is radiographically inspected—3-4 pictures are taken of each joint. For the latest orbiter, Challenger, only four of the tube welds failed to pass the rigorous inspection on the first try; 99.3% of the joints qualified.

The tubing, which transports such fluids as nitro-tetroxide, hydrogen, oxygen, and hydraulic fluid is made of various stainless steels—304L, 21-6-9, 15-5 PH, and 17-4 PH as well as Inconel 625 and 718 high-temperature nickel alloys. The sleeve material is 304L for the stainless tubing, matching Inconel alloys for the nickel-alloy tubing. Tube diameters

By Jerry Tryon and **Jay Contreras**
Senior welding engineers, Space Transportation & Systems Group, Rockwell International, Downey, Calif

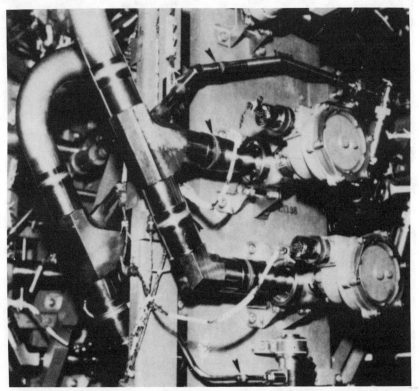

1. Arrows point to some of the tube welds in this section of the orbiter fuselage. Of the 550 required, most are made after much of the orbiter equipment is in place

2. One of the four modified Astro-Arc welding heads. Each is programmed to circumferentially weld stainless-steel and nickel-alloy tubing of various sizes

Simple tooling positions tube welds

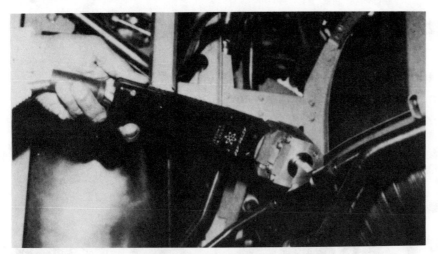

3. Welding head, closed around sleeved joint in congested fuselage, produces full-penetration welds in ½–2 min, depending on tube size

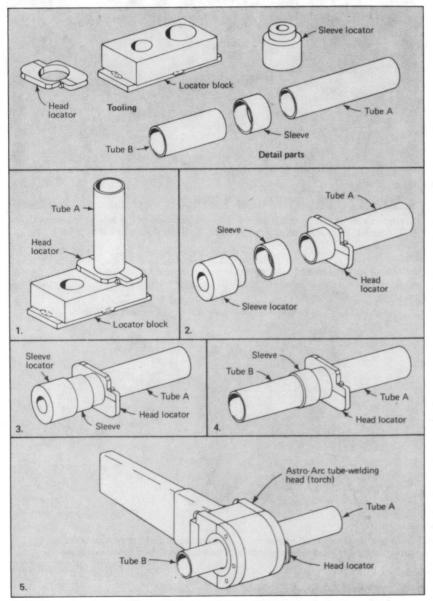

4. Tooling procedure positions welding head precisely around tube joint

range from ¼ to 2½ in.; wall thicknesses, from 0.020 to 0.042 in. Because of the various alloys and tube sizes, about 40 weld-processing schedules are required.

The welding operation is performed with modified Astromatic E-200-T clam-shell-type automatic tube-welders (Fig 2). The welding heads, which were modified to Rockwell specifications, are more compact than standard ones so that they can be used in the space-limited areas of the orbiter fuselage. Because of the different-size tubing, four heads were required. Placed around the sleeved joint areas (Fig 3) and closed by four attaching bolts, the heads are programmed to the 40-odd weld schedules.

The tooling concept that was developed ensures that the sleeve is directly centered around the butted tube ends and, also, that the welding-head electrode is directly in line with the tube ends. The procedure involves the use of locator blocks, welding-head locators, and sleeve locators, as shown in Fig 4.

First, for benchtop setups, the end of one of the tubes is placed in its appropriate recess in the locator block. The welding-head locator is placed over the tube against the locator block and tightened against the tube by means of a setscrew. For setups inside the orbiter fuselage, the welding-head locator is first slipped onto the tube.

The locator block is then removed, leaving the welding-head locator fixed firmly to the tube. With the help of the sleeve locator, the tight-fitting sleeve is placed on the tube. These locators are designed so that exactly one-half of the sleeve will be positioned on the tube end. The location of the edge of the sleeve is marked on the tube to designate its proper position in case it should be nudged out of place.

The other tube end is then inserted into the opposite end of the sleeve and butted against the end of the first tube.

The welding head is placed over the sleeved joint with one face of the head up against the welding-head locator. The electrode in the welding head will automatically be precisely in line with the butted tube ends.

The rest of the operation is routine. After an external and internal argon purge of the joint area, which takes 5 min or so, the joint is welded using a pulsed current of 20–100 A, depending on tube wall thickness. The weld is allowed to cool for about 5 min before the welding head is removed. Weld time is about ½–2 min per joint, depending on tube size.

The entire operation—sleeving the joint, assembling the welding head, purging, welding, cooling, and disassembling the head—typically takes 30 min. ∎

Reprinted from *Welding Journal*, December 1977

HARD TOOLING FOR THE FABRICATION OF THE EXTERNAL TANK OF THE SPACE SHUTTLE

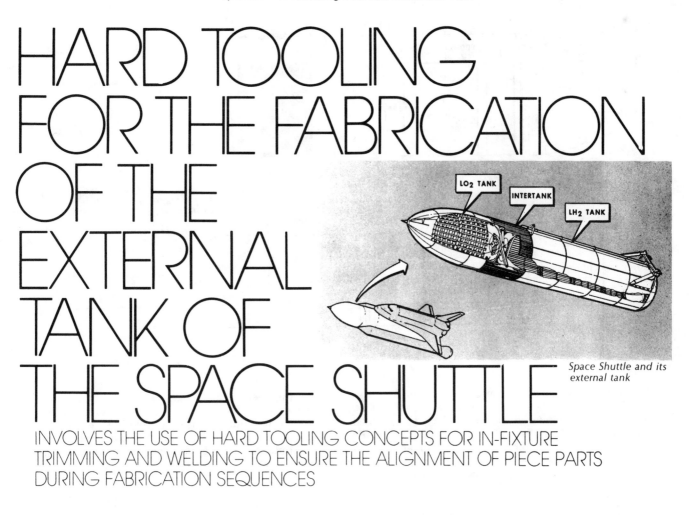

Space Shuttle and its external tank

INVOLVES THE USE OF HARD TOOLING CONCEPTS FOR IN-FIXTURE TRIMMING AND WELDING TO ENSURE THE ALIGNMENT OF PIECE PARTS DURING FABRICATION SEQUENCES

BY G. W. OYLER AND F. R. CLOVER, Jr.

Approximately three years ago, Martin Marietta received a contract to build the External Tank for the Space Shuttle at the Michoud Assembly Facility in New Orleans. Sponsored by NASA and directed by the Marshall Space Flight Center, the initial contract called for building three ground test and six flight test articles. To date, the first two test articles have been built. A follow-on contract may require the building of 439 External Tanks during approximately the next 15 years.

The Space Shuttle consists of an orbiter, an external tank, and two solid rocket boosters. After launch, the two solid rocket boosters which are attached to the two sides of the External Tank will be jettisoned at approximately 25 miles (40 km) and will be parachuted into the ocean, recovered, and recycled for reuse approximately 16 times. The External Tank, which supplies the fuel for

the orbiter, will be jettisoned in near orbit and break up as it re-enters the atmosphere; thus, an External Tank is required for each mission. The orbiter will perform its assigned mission, fly back and land on earth, and be recycled to be reused approximately 100 times.

The Space Shuttle, at launch, will weigh 4.4 million pounds (2×10^6 kg) and have a thrust capability of 6.8 million pounds (3.1×10^2 kg). The External Tank when empty weighs 75,000 pounds (340.2×10^2 kg) and when filled with liquid fuel weighs 1.638 million pounds (7.4×10^5 kg). The External Tank also serves as the backbone of the launch configuration as well as supplying the propellant for the orbiter portion of the Space Shuttle. The External Tank is 155 ft (47.2 m) long and 27.6 ft (8.41 m) in diameter. It is made up of two welded tanks connected by an intertank skirt.

The LH_2 tank will hold liquid hydrogen at a temperature of −423 F (−253 C); it is 96.7 ft (29.5 m) long and has a 27.6 ft (8.41 m) diameter. The LO_2 tank will contain liquid oxygen at −297 F (−183 C) and is 53.4 ft (16.3 m) long with a 27.6 ft (8.41 m) diameter. Both are shown schematically in Fig. 1.

G. W. OYLER—now Technical Director, American Welding Society, Miami, Florida—was Chief Welding Engineer and F. R. CLOVER, Jr., is Group Welding Engineer, Martin Marietta Corporation, Michoud Assembly Facility, New Orleans, Louisiana

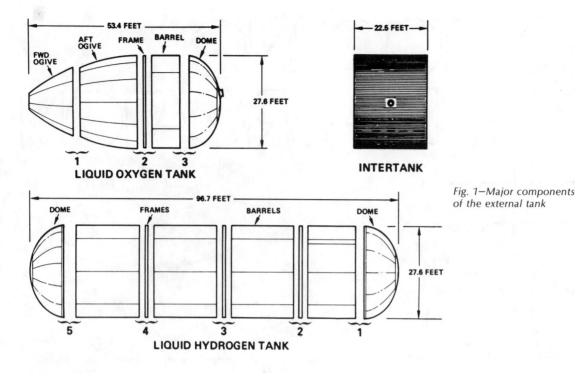

Fig. 1—Major components of the external tank

The two tanks are made from 2219-T87 aluminum with walls in the weld areas that, primarily, are 0.140 to 0.550 in. (3.6 to 14.0 mm) thick with selected reinforced portions as thick as 1.1 in. (27.9 mm). Each is fabricated by the welding of dome and barrel components to intermediate stiffening frames. The intertank skirt is made primarily from 7075-T6 aluminum and is 22.5 ft (6.86 m) long by 27.6 ft (8.41 m) in diameter. This skirt is fabricated using mechanical joining processes, and its purpose is to connect the two propellant-containing tanks to make up the so-called External Tank.

Tooling Concepts and Criteria

The tooling that has been designed and built for the fabrication of the various components—together with the assembly of the two tanks which make up the External Tank—is the subject of this paper. The concepts and criteria that were incorporated into the tooling are outlined in Table 1.

Since hundreds of these tanks are to be built, a hard tooling concept was utilized. These tools—or fixtures—are of rugged construction and are precision-built to provide consistency of all produced components for complete interchangeability.

Backing Bars

All metallic parts of the fixtures within 4 in. (101.6 mm) of the weld joint were fabricated from non-magnetic metals to eliminate the presence of any magnetic fields which would cause arc blow. The backing bars were made of Type 321 stainless steel and, generally, the supporting members were made from 6061 aluminum. No welding was permitted on the stainless steel backing bars or clamp fingers, since welding causes stainless steel to become magnetized. The backing bars actually consist of two bars spaced from ½ to 1 in. (12.7 to 25.4 mm) apart so that an observer can watch the forming of the penetration bead during welding and can accordingly

advise the operator to change the welding parameters.

All welds are of the free-fall type. A constant heat sink is essential to produce consistent welds from end to end. Therefore, the backing bars and the hold-down clamps were contoured to the configuration of the part. To provide positive contact, the edges to be welded are clamped.

Fixture Clamping

A closeup of the clamps used in the dome gore-to-gore fixture appears in Fig 2. These clamps are typical of those used on other fixtures. A separate clamping area at the bottom of the fixture accommodates a 2 ft (0.61 m) long preproduction plate. The clamps were made of Type 321 stainless steel and were contoured to conform to the configuration of the piece part. A clamping pressure of approximately 300 lb per linear inch (5.36 kg/mm) is exerted through the toe of the clamp onto the piece part. The segments of the clamps were aligned longitudinally within ±0.030 in. (0.76 mm) with a maximum spacing between the segments of 0.030 in. (0.76 mm).

To achieve positive contact, the surfaces of the backing bars and clamps were hand-polished to 32 RHR to assure essentially 100% contact to the material being welded. To prevent contamination of the weld, no paint,

---

Table 1—Tooling Concepts and Criteria

• Hard tooling	• Minimum offset
• Non-magnetic metals	• Precision trimming
• See-through capacity	• Minimum gap
• Constant heat sink	• Shrink fitting
• Positive clamping	• Precision seam following
• Clamp alignment	• Precision alignment
• Positive contact	• Vibration-free
• Contamination-free	• Integral radiography
• Positive positioning of parts	• Preproduction plate

Fig. 2—Hold-down clamps on dome gore-to-gore fixture

Fig. 3—Power saw in operation

grease, or oil was utilized on (or even near) the backing bars or clamping fingers. The clamps are actuated by air pressure using the fire-hose principle rather than using hydraulics to eliminate the risk of exposure to oil contamination.

To provide positive positioning of the parts, the entire periphery of the parts is held firmly in the fixture with up to 300 lb per linear inch (5.36 kg/mm) on the weld joint and with 150 lb per linear inch (2.68 kg/mm) on the remainder of the periphery. Locating and indexing holes are coordinated in the fixturing and trim areas of the piece parts to assure precision fitting in later stages of assembly. To assure minimum offset, the fixtures were precision built and aligned so that the offset between the two plates to be welded will be less than 0.010 in. (0.25 mm).

Trimming Operations

To provide precise trimming, the piece parts are firmly clamped in place by the fixture and are not released between trimming and welding. The straight-edged parts are trimmed to dimension using an air-driven power saw shown in Fig. 3. This air-driven power saw rotates at a speed of 3800 RPM. The saw blade was specially designed to be vibration-free and to provide an ultra-smooth joint edge for welding. The edges of both parts are cut using the same saw, operating on the same track to assure edge squareness within ± ½ deg and a maximum gap of 0.010 in. (0.25 mm).

Parts with curved edges are trimmed using an air-driven router. The router rotates at 22,000 RPM and is used for preparing the weld edges of curved parts such as fitting openings in the dome cap. A ½ in. (12.7 mm) diameter router bit is utilized to minimize vibration and chattering. All circular fittings are shrunk-fit with liquid

nitrogen; thus an interference fit is provided to eliminate the risks of weld cracking and distortion.

Referring again to Fig. 2, a minimum gap of less than 0.010 in. (0.25 mm) between the plate edges is assured by the precision trimming. In addition, straight-edged parts are held together by an inward force of at least 50 lb per linear inch (0.89 kg/mm) during welding. To provide precision seam following, the saw and welding heads were mounted and operate on the same track which is contoured to the part configuration. It is not necessary for the operator to make any positioning adjustments during welding.

Fixture Alignment

The fixtures were set up using a contoured master or splash gage and are checked periodically using the gage to assure alignment. The fixtures were initially aligned using optics and/or lasers to assure close tolerance alignment. To minimize vibration and movement, each tool was attached to a free-floating concrete slab which was anchored by numerous piles as long as 60 ft (18.3 m), driven into the earth. The operator rides in a conveyance on a separate track which is completely independent of the welding track and, therefore, the movement of the operator cannot cause any vibration of the welding head.

Many of the fixtures incorporate integral radiography so that the welds can be X-rayed and repaired without removing the piece part from the fixture.

Fixture Preproduction Plate

All fixtures have a separate portion for the making of a preproduction plate prior to the making of a production weld. This portion of the fixture, which accommodates a 2 ft (0.61 m) long straight plate, is identical to the fixture construction that is used for the actual piece part. This preproduction plate of the same thickness as the piece part is welded and visually inspected prior to the making of the production weld. This is done to assure that every parameter is under control and that an excellent quality production weld will be produced.

Welding Equipment

All welds are made using the gas-tungsten-arc process

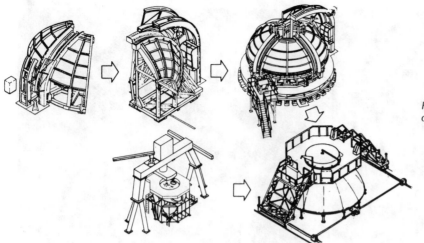

Fig. 4—Schematic representation of dome assembly sequence

with DCSP and helium gas shielding. The welds are square-butt and are made in three passes—that is, sealer or partial penetration pass, full penetration pass, and a filler metal pass. The arc length is controlled by a voltage-controlled head. All welds are either made in the vertical-up position or the downhand position.

In most cases, the welding torch travels on a propelled carriage. The tank assembly welds are an exception; here the tank is rotated and the torches are stationary. The welding current is controlled by using solid-state electronic power supplies.

Many of the weld joints are tapered in thickness and have as many as six different thicknesses within the length of one weld. To change the parameters accordingly, computerized taper controllers were designed and built to change the parameters with the accuracy that is required. Since the welds are of the free-fall type, drop-through must be controlled by the precision welding equipment and the consistency of the heat sink which is provided by the tooling.

Dome Fabrication

Figure 4 schematically depicts fabrication sequence of the dome, showing the five primary fixtures utilized in the welding of the piece parts which comprise the dome. Figure 5 illustrates these fixtures as they are arranged in the plant.

In order to minimize the quantity of large fixtures required, the three domes with somewhat different features were designed so that they could be built on the same fixtures. The dome subassemblies are fabricated using the ¼-panel technique on smaller, less expensive fixtures. The dome body is then assembled from these ¼ panels on a single fixture.

Gore-to-Gore Fabrication

In the fixture in the extreme upper right of Fig. 5 and upper left of Fig. 4, three gores are trimmed and welded to make a ¼ panel. The gores are stretch-formed, chem-milled, and coated with a protective coating by a vendor.

Fig. 5—Five dome assembly fixtures

Fig. 6—Dome body assembly fixture

Fig. 7—Vertical barrel assembly fixture

Fig. 8—Horizontal barrel assembly fixture

The protective coating is removed only from the area which is to be welded.

The fixture is made up of two halves that can be moved apart by traveling on air cushions. Gores are placed and clamped in each half of this fixture. The excess for trim, approximately ¼ in. (6.4 mm), of one edge of the gore in the left half of the fixture is sawed off. This half of the fixture is moved sideways. The other half of the fixture is then moved sideways so the gore is properly located under the saw and its excess is sawed off.

The same saw operating on the same track is used to trim both gores so that the cuts are parallel and perpendicular, thereby assuring a minimum gap of less than 0.010 in. (0.25 mm). The weld edges are then hand-filed and wire-brushed to provide the ultra clean edges required to produce high quality welds. The half of the fixture that was first moved is now moved back to the abutting position with the other half. The halves are pinned into position to provide the required alignment, an offset of less than 0.020 in. (0.51 mm). Pressure is then exerted transverse to the joint to provide an inward force of 50 lb per linear inch (0.89 kg/mm) during welding. The two gores are welded together.

The gores are not unclamped between trimming and welding in order to assure the best edge alignment. After welding, the two gores are unclamped and placed on the one half of the fixture and reclamped. The third gore is placed in the other half of the fixture and the steps just described are repeated to fabricate the ¼ panel.

In the second fixture from the upper right in Fig. 5, a stiffening frame is welded to the circumferential edge of the ¼ panel. This frame provides additional strengthening between the dome and the adjacent barrel. The ¼ panel is loaded in one half of the fixture and the frame is placed in the other half. Both are trimmed and welded similarly to that discussed for the gore-to-gore fixture.

Dome Body Assembly

The dome body assembly fixture is the third fixture from the right in Fig. 5 and is pictured by itself in Fig. 6. Four ¼ panels are inserted into respective sections of the fixture and clamped in place. Two ¼ panels are trimmed and welded to form half domes. The half domes, after careful sizing to obtain the proper circumference within 0.050 in. (1.3 mm), are trimmed and welded. Each ¼

segment of this tool can be moved outward in both the longitudinal and transverse directions to provide access for sawing and joint cleaning. The segments or nests are then brought together and pinned in position to provide the precision fit-up and alignment of the piece parts. The personnel carrier on its own track is visible at the upper center of Fig. 6.

Dome Cap

Essentially all of the openings, such as manhole access and fuel inlet and outlets, are in the dome cap. There are two fixtures like the one shown in the insert of Fig. 5 that are used for welding the fittings into the caps. This tool has its own machining head so that the diameter of the hole can be precisely routed to accept the previously machined fittings. The fittings are precision machined and the penetration opening sized to provide an interference fit. The fittings are immersed in liquid nitrogen, then removed, and inserted into the opening in the dome cap. Hold-down clamp tooling is installed to provide the alignment and heat sink. The piece parts are brought back to room temperature by the use of heat lamps which are located on the underside of the backing bars, and then welded.

The dome cap-to-dome body weld is made in the fixture on the left in Fig. 5. This fixture also has a precision machining set-up that routs the circumference of the opening in the dome to a close tolerance. The dome cap is machined on a separate position of the fixture so that the gap between the dome body and the cap is no more than 0.010 in. (0.25 mm). The dome cap

Fig. 9—Forward and aft ogive assembly fixtures

Fig. 10—LO₂ tank assembly fixture

and the fixturing which restrained it during machining are then lifted up and placed into the dome body opening and the cap is welded into the dome.

Barrel Fabrication

The barrels are of three different lengths—8, 15 and 20 ft (2.4, 4.6 and 6.1 m). The shorter two barrels are made in the vertical welding fixture shown in Fig. 7. The 15 ft (4.6 m) long barrel is made from 8 panels and two longerons, while the 8 ft (2.4 m) long barrel consists of 4 panels.

The barrel panels are received from a vendor and routed to the proper width on a separate table. These flat panels are then draped over a curved table shown in the foreground and restraining devices are attached to both ends of the panels and inserted into the barrel fixture. The joint edges of two panels are then welded and the operation is repeated nine times to complete the 15 ft (4.6 m) long barrel. The welded panels are rotated 90 deg in the fixture and the weld is X-rayed. (The welds which were made during the dome fabrication were X-rayed on a separate fixture after the dome had been completed). The completed barrel is then removed by a crane.

The 20 ft (6.1 m) long barrels, containing 8 panels of which there are three per LH₂ tank, could not be built in the vertical position because the crane hook height in this building was not sufficient to permit removal of the completed barrel from the fixture. Thus, the long barrels are fabricated in a horizontal welding fixture (sometimes called a "cigarette roller") which is pictured in Fig. 8.

Flat sheets are introduced into the fixture by means of the feed table shown in the left foreground of Fig. 8. The table is also the routing station for sizing the panels and for weld edge preparation. Gripping clamps are attached to the ends of the plate and the plate is pulled partially around the periphery of the fixture and clamped for welding at the weld station. A second flat plate is similarly introduced and its edge is also routed.

The weld is made in the downhand position from the inside-out. The two panels are then pulled further around the periphery of the fixture and the operations are repeated seven times until a complete barrel is

fabricated. The welds are X-rayed in this fixture at a position 45 deg from where the welds are made. After completion of the barrel welding, a special removal device is utilized to unload the barrel from the fixture. The welding fixture had to be constructed using the cantilever principle to provide clearance for the removal of the barrel from its fixture.

Ogive Fabrication

The LO₂ tank has an ogive or bullet-like front portion. This ogive portion is 32 ft (9.8 m) long and must be made in two sections since the stretch-forming vendor could not stretch-form the full length panels. The fixture on the left in Fig. 9 is used to make the forward ogive section, and the one to the right is utilized to make the aft ogive section.

The forward ogive is made up of eight gores which are progressively clamped, sawed, and welded to form the component. An end or nose fitting is also welded into the forward ogive using the clamp assembly at the top of the fixture. The nose fitting weld is the only weld in the fabrication of the external tank that is made in the horizontal position. The ogive section is then removed and X-rayed in a separate fixture.

The aft ogive is made from 12 gore panels that are progressively clamped, sawed, and welded in the fixture. The welding of all the ogive gore panels is accomplished in the vertical-up position. Again, the component is removed and X-rayed in a separate fixture.

LO₂ Tank Assembly

To fabricate the 53.4 ft (16.3 m) long LO₂ tank, four circumferential welds must be made. The fixture shown in Fig. 10 is designed so that it rotates past the trim saw or router and the welding torch. The saw and routing station is shown in the lower center of the photograph. The head stock and the tail stock of the fixture can be synchronously driven or independently driven, as required. The welding heads are located near the top of the fixture at 1:30 o'clock on a gantry (Fig. 11*) which moves longitudinally to the rotating fixture.

To make the forward ogive-to-aft ogive weld which is at a smaller diameter, a portion of the gantry is hinged so that the welding torch can be placed at the proper diametrical position and the weld is made by rotating the tank past the stationary torch. The two components

---

*Although the gantry in Fig. 11 is actually used for LH₂ tank assembly, it is identical to the gantry used for the LO₂ tank. The welding gantry is lead-lined to provide shielding during radiographing of the completed weld.

Fig. 11—LH₂ tank assembly fixture

Fig. 12—Internal expanding mandrel

are held in place by an internal expanding mandrel shown in Fig. 12. The mandrel is expanded to provide the required fit-up, a maximum offset of 0.020 in. (0.51 mm), and the weld is made.

The next operation is to attach the ogive to the barrel; however, a T-ring or stiffening ring approximately 5 in. (127 mm) long must be inserted between these two components. This ring is machined to dimension on a large boring mill prior to being inserted into the fixture. The ogive and barrel edges are sawed to the proper length. The three components—the ogive, T-ring and barrel—are brought together and held in place by two expanding mandrels. The two welds which are only some 5 in. (127 mm) apart are made simultaneously using two welding torches. These welds are made concurrently to prevent bell-mouthing that would occur if made separately; it also reduces the welding time by 50%.

The last weld made in the LO₂ tank assembly fixture is to join the barrel to the dome. Prior to placing the dome on the fixture, a complicated slosh baffle must be

inserted into the ogive-barrel subassembly. Also, a close-out mandrel shown in Fig. 13 is attached to the stiffening frame of the dome. The dome is then placed into the fixture and the barrel and dome edges are sawed to size. The mandrel is expanded to provide the required fit-up and the weld is made.

Fig. 13—Closeout expanding mandrel

Fig. 14—LO₂ and LH₂ tank assembly fixtures

The welds are X-rayed by placing the film on the outside diameter of the welds and exposing the film by use of a radioactive source extended into the center of the tank.

LH₂ Tank Assembly

A fixture similar to the one just discussed is used to assemble the 96.7 ft (29.5 m) long LH₂ tank. This fixture is 160 ft (48.8 m) long and 32 ft (9.75 m) in diameter.

The components consist of two domes, four barrels, and three T-rings; these are welded together much the same way as described for the LO₂ tank. The dual-torch principle is used for making three sets of welds in this tank. The same types of mandrels are used for welding both tanks.

Expanding Mandrels

Figure 12 depicts one of the expanding mandrels that is utilized to form the circumference of the piece parts so that the offset is less than 0.020 in. (0.51 mm). After being inserted into the tank on a boom extension, the mandrel is expanded using pressures up to 90 psi (620 × 10³ Pa) to form the two parts to obtain the desired alignment.

A close-up of a segment of the expanding mandrel is shown in the lower portion of Fig. 12. To have the mandrel extend to its full diameter, one portion is first extended and then a second portion is extended and locked into position to form the full diameter. The pressure on each segment, which is generated by the fire-hose principle, is progressively increased until the minimum offset is obtained.

In those cases where the two welds are made simulta-neously, two mandrels are required—one located under each weld joint. The mandrel is retracted by first releasing the air pressure in the fire hose and then retracting the locking segments and the other segments.

Figure 13 shows the mandrel and a closeup of a portion of the mandrel that is used for the closure welds of each tank—that is, the dome to the last barrel section. The segmented mandrel is first attached to the stiffening frame of the dome as shown in the upper portion of Fig. 13 before the dome is placed into the tank assembly fixture. The mandrel can exert pressures up to 75 psi (517 × 10³ Pa) to make the parts conform to the desired configuration and to provide the minimum offset. After the joint is welded, the mandrel is disassembled into hand-carryable segments, 40 lb (18.14 kg) each, and removed from the barrel through the access opening in the dome of each tank.

Figure 14 shows the location of the two fixtures that are used for assembly of the LO₂ and LH₂ tanks. The LO₂ tank on the right is completed and ready for removal from the fixture. The LH₂ tank on the left is nearly completed with only the closure dome to be welded in place.

Summary

The fixtures which are used for making the External Tanks are really precision machines which must hold close tolerances throughout the life of their expected usage. It is expected that as many as 439 of these External Tanks will be built during the next 15 years. At the present time, there is a total of 15 major weld fixtures now in use for welding the External Tanks.

The Space Shuttle is expected to undergo its first test flight in March of 1979 with its first mission being scheduled in May of 1980.

CHAPTER 5

MANUFACTURING SYSTEMS

NASA's Use of Robotics in Building the Space Shuttle

by K. Fernandez,
C.S. Jones, III,
and M.L. Roberts
NASA

INTRODUCTION

There are many benefits that have come from the U.S. Space program. Some benefits like satellite communications are highly visible in our daily lives. An area of technology often overlooked is NASA's contributions to the development of materials and manufacturing processes. The Materials and Processes Laboratory (M&P) at MSFC under the directorship of R.J. Schwinghamer has the responsibility for developing materials and manufacturing processes that are used by NASA. This paper will describe three areas in which the M&P Lab has applied robotics to improve reliability and performance of Shuttle components, these areas include: 1) three thermal protection spray systems; 2) one thermal protection removal system; and 3) two arc welding systems under development. The six robotic systems described in this paper increase in complexity from a basic spray system to a welding system that will be used for research in sensor technology.

Shuttle Configuration Overview

Before examining the areas of robotic activity it may be useful to briefly review the Shuttle vehicle and its launch components. Unlike previous NASA vehicles such as the Mercury, Gemini, and Apollo series that used totally expendable launch systems the Shuttle and most of its elements are completely reusable. Figure 1 shows the Shuttle shortly after launch with the External Tank (ET) and Solid Rocket Boosters (SRB's) attached. Thrust for lift-off is provided by the SRB's and the three Main Engines aboard the Orbiter. Fuel (liquid hydrogen and oxygen) for the Main Engines is supplied by the large External Tank. After the SRB's are burned-out they are jettisoned and recovered for re-use on a future launch. When the Orbiter has reached the appropriate altitude and velocity the Main Engines are cut-off and the External Tank is jettisoned. The External Tank is the only element of the Shuttle's launch cluster that is not recovered at the present time. The great cost reductions per mission that have been achieved by the Shuttle are due in part to its re-useability and to its land-based recovery.

Thermal Protection Systems

Thermal protection systems are an integral part of any space vehicle and the Shuttle is no exception. The Orbiter itself is equipped with a covering of insulating tiles to protect it from the heat of re-entry. The SRB's and the ET must also be protected: the ET though expendable must be insulated to reduce boil-off of its cryogenic contents and to reduce the accumulation of ice prior to launch; and the SRB's must be protected during their re-entry if they are to be used again. The protective coating developed by the M&P Lab for the SRB's is called the Marshall Sprayable Ablative or MSA. MSA is basically an epoxy-terminated urethane resin filled with various organic fibrous ingredients for strength. The result is a low density ablative with the necessary insulating properties. A second material formulated for use on the External Tank is called SOFI (Spray-On Foam Insulation) which is a lower density urethane material with a Freon blowing agent. SOFI is designed mainly for insulation rather than ablative purposes.

ROBOTIC SPRAY SYSTEMS:

MSA and SOFI are applied by spraying, however, the protective clothing required and the dimension of the workpieces (12ft diameter for the SRB and 30ft diameter for the ET) make manual spraying virtually impossible. The solution developed by the M&P Lab involves the use of a rotary table for positioning and rotating cylindrical sections of the SRB's and the ET and an industrial robot for positioning the spray gun. There have been three robotic spray systems developed. Figure 2 depicts the general appearance for all three facilities. The robot and turntable are housed in a spray cell with the air temperature and humidity closely controlled. Facility personnel view the spraying operations and monitor the instruments from the control room immediately adjacent to the spray cell. The differences between these systems are in the level of complexity of the robot and its interaction with other system components. We will now describe these systems.

System 1: Solid Rocket Booster Spray Development System

The first MSA spray facility was assembled at the M&P Laboratory in 1978 to apply the MSA to the Solid Rocket Booster. This facility incorporates a Trallfa T2000 robot, an electrically driven rotary table, and a Digital Equipment Corporation 11/23 mini-computer. Figure 3 depicts the functional block diagram of this system. The robot's spray path is pre-stored in the T2000 using the robot's continuous path programming feature. During spraying operations the turntable speed is set to develop the desired spray surface feed rate. Robot motion is initiated by a signal pulse from the rotary table when a reference position is encountered. The DEC 11/23 computer is used primarily as a data acquisition system, data-logging device, and to calculate table rotation velocities needed for a specified coating thickness. This system has been used primarily to determine the optimum conditions for applying the MSA.

System 2: Solid Rocket Booster Spray System

The second MSA spray facility was implemented by United Space Boosters Incorporated (USBI) at the Kennedy Space Center (KSC) in 1980 after favorable results were obtained on the MSFC system. This system again utilized the Trallfa robot, however the robot was mounted on both rails and a hydraulic lift-table. Although the motion of the lift-table and roller was not coordinated with the robot it did allow a much larger item to be sprayed in several stages. One additional variation of this system is that the table is used only for positioning a sector of the cylindrical spray surface, while the Marshall system 1 actually used the table's rotary motion during spraying operations. Three working cells identical to system 2 are currently used at KSC to spray MSA on the Solid Rocket Booster elements prior to launch.

System 3: External Tank SOFI Development System

The third spray facility was again developed at the Marshall Center's M&P Laboratory. The purpose of this system is to further test the SOFI spray process for a wide range of spray parameters. Since this system is designed for experimental study of SOFI as applied to the 30 foot diameter External Tank a higher surface feed rate and spray flow rate had to be achieved and consequently a robot with a larger work envelope was required. The resulting system uses a Cincinnati Milacron T3 robot, a larger Rolair air bearing rotary table, and a Digital Equipment Corporation 11/23 computer as its main components. The functional block diagram for this system is shown in Figure 4. Although this system shares similarities with systems 1 and 2, the primary differences are in the system's physical capacity and an increased role for the external DEC 11/23 computer. In addition to data-logging the 11/23 is used to up and down-load the T3's controller via the T3's RS232 data link. This feature greatly facilitates the large number of production runs made during the M&P Laboratory's spray development program.

System 4: MSA Removal Facility

The next robotic facility to be described is currently being checked-out at the M&P Lab. This system is used to develop a method for removing the MSA material from the SRB's after their post-launch recovery. Removal of the partially ablated MSA is necessary prior to re-spraying. One method found to be very effective uses a narrow high pressure (20000 psi) stream of water to literally cut away the old MSA. This method of removal uses a "Hydro-laser" designed by the Tritan Corporation.

The system configuration of the MSA removal system is shown in Figure 5. A Cincinnati Milacron HT3 robot is used to hold the "Hydro-laser." The robot is mounted on a hydraulic lift table similar to that used on system 2 described above, however the lift table's height is sensed by a position encoder and fed to the 11/23 computer. The 11/23 then uses this robot height data to adjust the programs that are transmitted to the Cincinnati Milacron's controller via the RS232 link. The DEC 11/23 computer is also used to monitor the safe operation of the pumping equipment, infrared proximity sensors for collision avoidance, and other system parameters. Should an unsafe condition be detected the computer will issue abort commands to the individual system components in the proper sequence to insure a safe system shut-down. Future enhancements planned for this system include adding a vision system that will be used to detect areas where the "Hydro-laser" failed to remove the MSA. Position information of these MSA residues in sensor coordinates along with table position data will be used to automatically program the HT3 to perform spot removal.

EXPERIMENTAL WELDING SYSTEMS

The final area in which the M&P Lab has applied robotics is arc welding. Two systems, one for the External Tank and the other for the SSME, are currently being developed. The details of these systems are given below.

System 5: External Tank Welder

The External Tank built by the Martin Corporation is constructed mainly of aluminum using the TIG welding process. At present nearly all welding is automated using an elaborate tooling fixture. If no problems occur this method is very efficient, but when a weld must be re-worked this specialized tooling is tied up while the repairs are made delaying the production of additional units. One solution to this problem is to off-load the unit to be repaired and perform the repair operation with a robot in place of the tooling. M&P plans to use a Cincinnati Milacron T3-746 robot equipped with both TIG and Plasma welding equipment for this operation.

System 6: Engine Welding System

The three Space Shuttle Main Engines (SSME) mounted directly on the Shuttle are manufactured by the Rocketdyne Division of Rockwell International in Canoga Park, California. A total of twenty-seven flight engines have been ordered to date with a production rate of six engines per year. The SSME is constructed primarily of high temperature stainless steel and welded using the TIG and EB processes. A picture of the engine assembly is shown in Figure 6 giving an indication of its complexity. Currently more than sixty percent of the SSME welds have been automated by Rocketdyne using hard tooling, however the automation of the remaining welds would not be cost-effective using these techniques. Attempts to use existing robots and sensors are hampered by the wide variation of weld types, weld positions, access, and fit-up.

Working in support of our engine manufacturer the M&P Lab is developing an experimental robotic welding system to automate the remaining manual welds. The Metal Processes Branch of M&P is at this writing formulating its requirements for this system. A functional block diagram is given in Figure 7. Since selections have not been made no actual hardware is specified, however the system structure is correctly shown.

The robot will contain a minimum of five degrees-of-freedom and work in conjunction with a two degree-of-freedom parts positioner. In addition to controlling robot and positioner motions the robot's controller will also control the TIG welding power supply. The system described thus far is available as an off-the-shelf item from several manufacturers, where M&P's plans depart from standard equipment is in the use of an external computer to support sensor experimentation and development. Receiving torch position data in world coordinates from the robot's controller and error data from the sensor the external computer will perform the necessary transformations and issue correction commands to the robot. Experimentation with a variety of non-contacting sensors is planned including vision systems, IR detectors and arc voltage sensors. In cooperation with other agencies and private industry NASA hopes to further develop welding sensor technology.

FUTURE DEVELOPMENT PLANS:

To aid NASA's future use of robotic systems and to enhance current system development the M & P Lab is developing a robot graphic simulator. This simulator will use a computer graphics display terminal to generate the kinematics of our various robotic systems allowing the rapid investigation of various robot and tooling configurations without the expense of actual system fabrication. Eventually this system will also be used to verify robot path programs that have been generated off-line.

CONCLUSIONS

The program of robotic systems use and development at the Marshall Space Flight Center's Materials and Processes Lab has evolved from applications to research in computer aided manufacturing. The goals of this program are primarily to improve productivity within NASA's programs, however the technology developed will undoubtedly benefit the private sector as well.

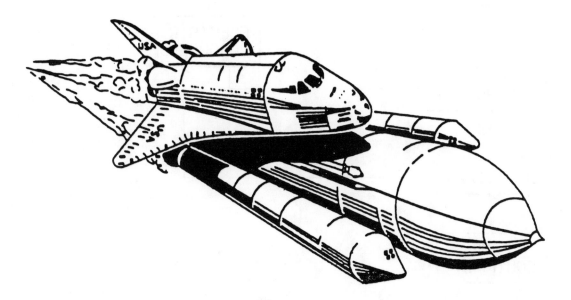

FIGURE 1
SHUTTLE LAUNCH CONFIGURATION

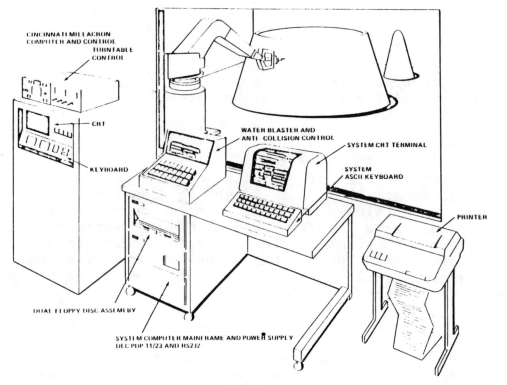

FIGURE 2
TYPICAL ROBOTIC SPRAY CELL

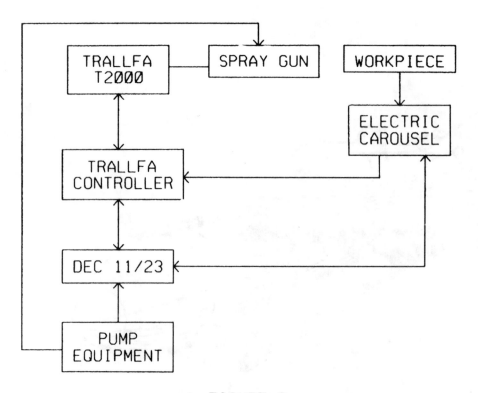

FIGURE 3
SPRAY SYSTEM 1 & 2

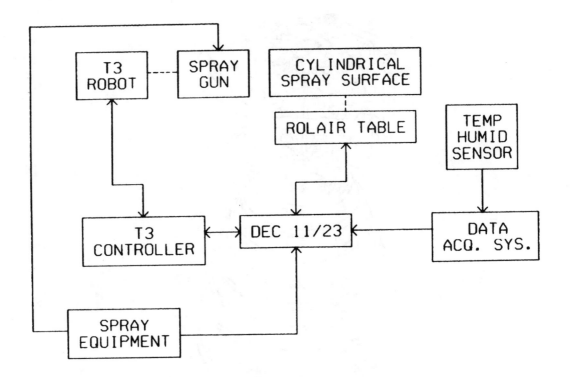

FIGURE 4
SPRAY SYSTEM 3

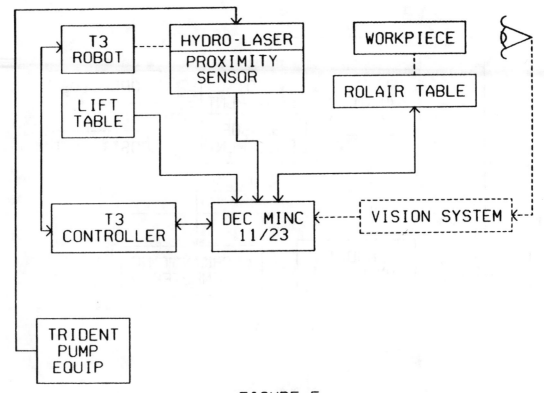

FIGURE 5
MSA REMOVAL FACILITY

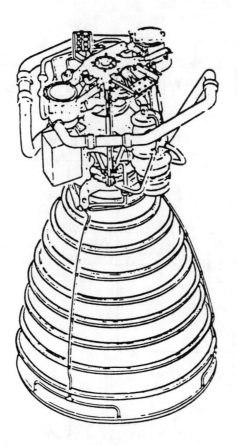

**FIGURE 6
SPACE SHUTTLE MAIN ENGINE**

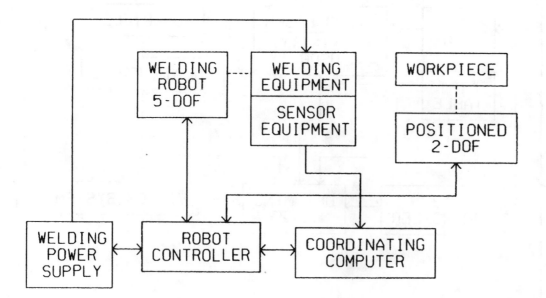

FIGURE 7
SSME WELDING SYSTEM

Presented at the RI/SME Applying Robotics in the Aerospace Industry Conference, March 1984

Robotic Processes Geared Towards the Space Shuttle External Tank Manufacturing

by Carlos A. Ramirez
Martin Marietta Aerospace-Michoud Division

Figure 1 shows the major components of the Space Shuttle, which are the Orbiter, the Solid Rocket Boosters (SRB's) and the External Tank (ET). The thrust for lift off is provided by the SRB's and the three Main Engines aboard the Orbiter. The ET supplies the fuel (liquid hydrogen LH2 and liquid oxygen LO2) for the Main Engines.

Martin Marietta Aerospace - Michoud Division is the prime contractor for the ET. The major components of the ET are the LO2 tank, the Intertank (IT) and the LH2 tank.

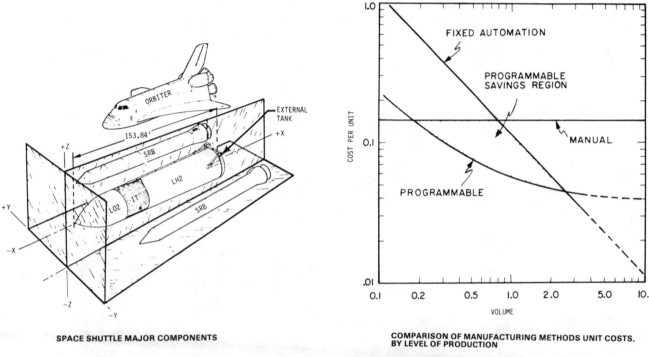

SPACE SHUTTLE MAJOR COMPONENTS

Figure 1

COMPARISON OF MANUFACTURING METHODS UNIT COSTS, BY LEVEL OF PRODUCTION

Figure 2

In 1981, NASA installed two Cincinnati Milacron HT3 robots (T3-586's) at the Michoud Assembly Facilities (MAF).

The large size of the ET subassemblies, their high dollar value and the relatively low manufacturing volume were the major constraints for turn key type robotic applications at MAF. P. Lynch, in his doctoral dissertation at MIT (1), shows that robotic applications are economically more attractive when the production volume is neither extremely low nor extremely high. This model is depicted in Figure 2.

Laboratory development and a payoff technology assessment study (PTAS) are ongoing efforts being made to apply robotics for the manufacture of the ET.

In manufacturing-processes-development, the following experiments were conducted and evaluated:

o Machining of super light ablator (SLA)
o Permanent fastening using a portable driller riveter
o Drilling through bushings
o Measurement of a gore panel thickness
o Ablator spray

In the area of technological development, the following experiments were conducted and evaluated:

o Sensory feedback for robot positioning enhancement
o Robot self-programming
o Off-line programming

In addition, experiments for inspection with vision using an Automatix II system and development of enhanced safety systems are in progress.

MANUFACTURING PROCESSES

MACHINING OF SUPER LIGHT ABLATOR (SLA)

The Thermal Protection System (TPS) is primarily utilized to maintain the LO2 and LH2 boiloff rates below the vent valves capabilities, to minimize ice formation on the ET surfaces, and to assure a low altitude fragmentation to meet the 100 x 600 nmi foot print limits (2). The super light ablator (SLA-561) is a highly filled elastomeric material that constitutes part of the TPS.

Robotic SLA machining was evaluated using 24" x 24" x 1" ablator panel glued to an aluminum substrate. The robot picked the substrate with SLA and placed it on a machining fixture. The machining end effector consisted of a 2.5 HP AC motor and a 4" diameter 60 grit grinder cup attached to the shaft of the motor. For debris pick up, a vacuum hose was attached to the cutter guard as shown in Figure 3a. It was found that the high thermal resistance of SLA inhibited heat dissipation, causing scorching on the SLA surface. Three air jets were installed to improve cooling of the grinder.

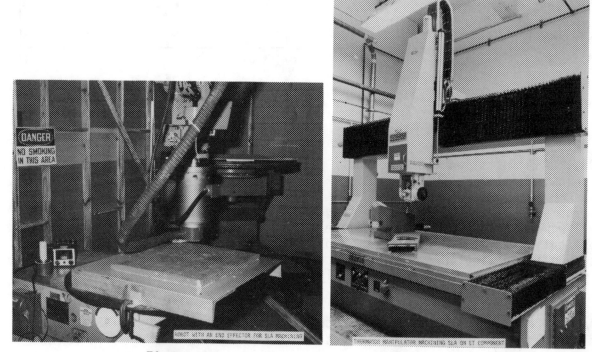

Figure 3a   Figure 3b

Factorial tests for three levels were run. The factors were depth of cut (0.02", 0.1", and 0.25"), feed rate (.5 IPS, 1.0 IPS, 3.0 IPS) and rotational speed (3867 RPM, 2906 RPM, 2012 RPM).

Aspects evaluated were the smoothness of the surface, the cork exposure and the dimensional tolerances.

Feed rates greater than 3 IPS tore the material. The maximum depth of the cut was limited by the cutter guard and the power of the motor. Measurements to size the motor show that to cut 0.25" at 3 IPS, a 7.5 HP motor would be needed.

The best combination of RPM and feed rate for the following depths of cut were: 0.02" depth of cut, 0.5 IPS, 3867 RPM; 0.1" depth of cut, 1.0 IPS, 2012 RPM; and 0.25" depth of cut, 1.0 IPS, 2906 RPM.

Tests with a metallic cutter indicated that the optimum cutter should be metallic for fast material removal with a grinding material attached to the base for surface smoothing.

On the production floor two Therwood cartesion-5 manipulators, shown in Figure 3b, were installed to perform the machining of SLA covered components.

PERMANENT FASTENING

A novel approach for permanent fastening was tested. It consisted of a close clearance GEMCOR portable driller/riveter attached to the robot arm. This configuration has a great potential for edge riveting on difficult access structures (Figure 4). The GEMCOR weighs 206 lbs., and can place 1 rivet in 5 seconds. The upsetting force of this unit is 6000 lbs. and the rivets (universal head of countersink 3/32" shank and 15/32" length) are fed automatically from a hopper to the GEMCOR head through a transparent flexible tube. This system installed the rivets with ± 0.012" accuracy.

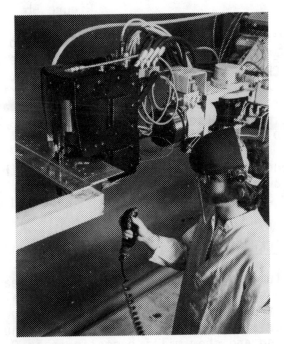

Figure 4 - Gemcor Portable Driller Riveter attached to an HT3 robot.

Caution must be exercised to program the robot with the GEMCOR, due to the overshoots caused by the size of the load. The bigger the mass inertia, the further apart the poles along the vertical imaginary axis in the Nyguist diagram (3).

DRILLING THROUGH BUSHINGS

A drilling end effector, shown in Figure 5a, consisted of a carriage that moved up and down through two precision ground rods. This carriage held a quakenbush drill motor mounted on a remote center (RCC) compliance mechanism to compensate for errors in positioning.

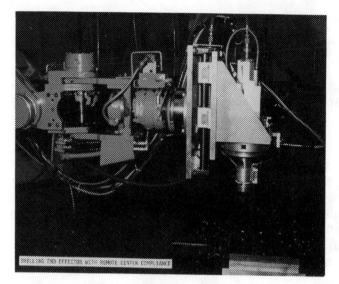

Figure 5a

Figure 5b

Two linear variable differential transducers (LVDT) were placed in quadrature, to sense the deviation of the remote center compliance (RCC) mechanism. The LVDT signals were displayed on a CRT to aid programming of the robot (Figure 5b). Switches detected carriage up or down position, drill in bushing, and drill cycle complete. The robot controlled the start of the drill cycle, the coolant and the detraction of the drill. The drill cycle was 14 seconds when drilling a 3/4" aluminum plate with 0.25" titanium nitride plated drill bit.

The robot carried the end effector to a drilling plate with 12 holes. Six holes had 0° chamfer, 3 had 20° chamfer and 3 had 45° chamfer. Out of 2700 drill cycles the robot missed only 3 holes. The holes were inspected with a go/no-go gauge (0.2495/0.2525").

MEASUREMENT OF GORE PANEL THICKNESS

The thickness of E.T. aluminum gore panels are manually measured at about 1" intervals. Then an elevation or thickness contour/map is manually drawn. Then, the gores are covered with a transparent material and, starting with the thickest regions, the transparent material is scribed and peeled off manually for a chem-milling process. This scribing and chem-milling continues in 0.003" increments until the specified thickness is obtained. Some of these gore panels are shown in Figure 6a.

A demonstration of the feasibility to automate this labor intensive process with a robot was implemented. The system, shown in Figure 6b, consisted of an ultrasonic sensor held by the robot. The robot commanded a DEC 11/24 computer to read the thickness signal of the ultrasonic transducer. After the robot scanned and measured the thickness all along and across the aluminum panels, the computer arranged the thickness data and built a data base for contour mapping. On the production floor, the robot can be mounted with a pen to draw the map on the surface. Then the gore panel can be covered with a transparent material. And finally the robot, with a scribing end effector, can perform the cutting of the protecting film for ulterior chem-milling.

Figure 6a

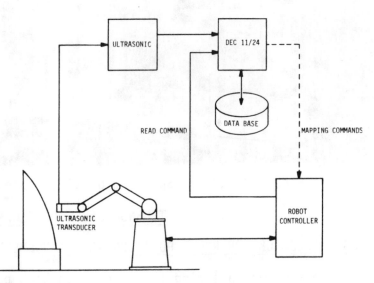

BLOCK DIAGRAM FOR ALUMINUM PANEL THICKNESS MEASUREMENT AND MAPPING FOR CHEM MILLING

Figure 6b

One of Martin Marietta's major subcontractors is evaluating the implementation of Gantry robots for Gore Panel Thickness Measurement on the production floor.

ABLATOR SPRAY

The spraying of SLA takes place in a hazardous environment (Class I, Division I). This is a tedious labor intensive process in which layers of about 0.005" are sprayed at 5 minute intervals until the sprayed thickness builds to about 0.75".

At present, efforts are concentrated to develop SLA spray on ET components using at HT3 robot. An end effector with a pneumatic actuator, which flips the gun 90° under the control of the robot, will be utilized. The robot spraying on ET component mock-ups is shown in Figure 7.

The lack of dexterity of the hydraulic robot for spraying was overcome by the additional degree of freedom provided by the actuator that holds the gun, as well as by the positioning table that holds the parts to be sprayed.

A non-hazardous material was developed to evaluate the effectiveness of spraying with the HT3 robot. This was done prior to incurring the expense of installing the robot in a Class I, Division I area.

The planned date to start production operation using this system is by the end of 1984.

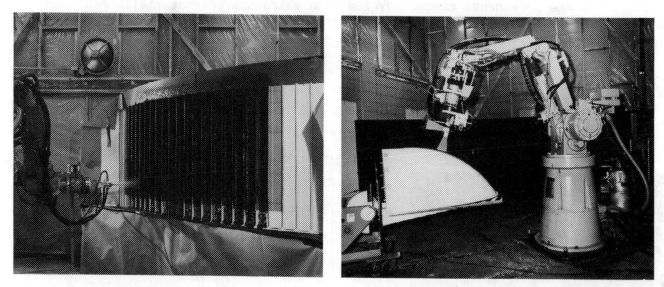

Figure 7 - Robot Spraying on ET Component Mock-ups

TECHNOLOGICAL DEVELOPMENT

SENSORY FEEDBACK FOR ROBOT POSITIONING ENHANCEMENT

There are manufacturing tasks in which high payloads and tight accuracies are needed. A benefit of a hydraulic robot like the HT3 is the high payload capacity (225 lb. with a CG at 10" from roll plate) however, its accuracy (±0.050") is not as good as the electric robots for tight tolerance manufacturing.

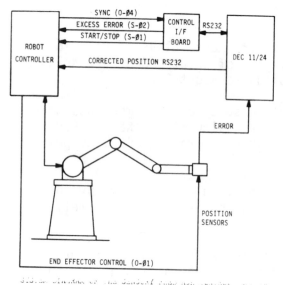

Figure 8a

Figure 8b

Real time sensory feedback control enhanced the positioning accuracy of the HT3 robot from ±0.050" to ±0.005". A sensory feedback control system was developed and demonstrated. The position accuracy of the HT3 robot was enhanced from ±0.050" to ±0.005". This was achieved using sensors that measure the error between the desired position and the actual robot position. These sensors are sampled by an analog I/O board of a DEC 11/24 mini-computer which calculates the magnitude and orientation of the error.

The mini-computer calculates the new corrected coordinates and commands the robot to the desired position. A block diagram of this system is shown in Figure 8a. Figure 8b shows the robot with the tactile accuracy end effector.

ROBOT SELF PROGRAMMING

Programming of the robot to perform manufacturing tasks (i.e., spraying, machining) on complicated contoured shapes is tedious, time consuming and hazardous to the operator since he has to get close to the robot and the part to ensure correct robot positions. Distance from the part surface to the tool held by the robot as well as normality to the contoured surfaces must be controlled.

Artificial intelligence for robot self-programming was demonstrated on contoured surfaces. The robot learns the characteristics of a contoured surface on a rough quick pass across such a surface. As the robot learns the characteristics of the curved surface, an algorithm that runs on a DEC 11/24 computer calculates the coordinates the robot must have to keep a constant distance away from and normal to the contoured surface. The systems block diagram is depicted in Figure 9a.

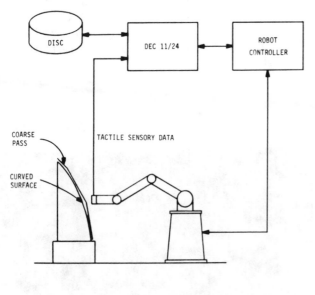

SYSTEM DIAGRAM FOR ROBOT SELF-PROGRAMMING

Figure 9a

Figure 9b

The robot arm had a three finger end effector. These fingers are linear potentiometers (LP's) (Figure 9b) which are used to determine a vector normal to the surface and the distance from the robot tool tip to the contoured surface. For demonstration purposes, a gore panel of the Space Shuttle External Tank was used.

OFF-LINE PROGRAMMING

Off-line programming for the Cincinnati Milacron T3 and HT3 robot was developed and demonstrated.

Off-line programming software was written in basic and the communication between the robot and the mini-computer Z89 was via a RS-232C line. The software is conversational and prompts the operator on the CRT screen with questions to aid him or her to program the robot. The operator is asked for:

(a) Coordinates of the points (path) the robot will move as well as the robot arm orientation,
(b) The function to be performed at each point,
(c) The tools to be activated,
(d) The I/O lines to be read or to be activated to synchronize operation with other automatic machines,
(e) The interrupt signals to which the robot must be sensitive for added flexibility in the programming and for safety implementation.

Once the sequences/routines have been created/edited, they are downloaded to the robot using a RS-232C interface. The transmission rate could be 600,1200, 2400, 4800 and 9600 BPS. An enhanced version of this program can provide pictorial images on the CRT of a CAD/CAM system showing the dynamics of the robot in action.

VISION SYSTEMS

A high-speed-image-data interface between a Gould SEL 32/27 was developed. This interface, which is shown in Figure 10, accepts image data from any device with RS 170 or NTSC format (e.g., cameras, videos, thermographer) at 6 MBS for real time processing. In addition, an Automatix vision system interacts with the Gould processor for distributed image processing. This system is being utilized to evaluate the use of thermographer in welding, to measure peaking and mismatch (Figure 11) of welded plates, alignment of large structures.

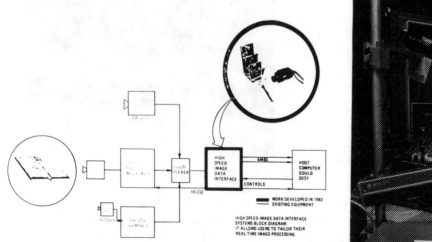

Figure 10

Figure 11

The use of a structure light using a HeNe laser and narrow band filters promise a lot of potential for implementing the aforementioned applications (Figure 10).

INTELLIGENT SAFETY SYSTEMS

A common practice to insure safety is to limit personnel access to the robot working area by physical (chains, walls) or electronic means (light walls, pressure mats, sensors).

In the manufacturing of the E.T. the concern with regard to safety is not only personnel, but also hardware. This is because of the expense of the parts and assemblies with with the robot interacts.

The measurements used to determine the stopping distance after detection of failure show that the arm would travel 3.5" after an emergency hydraulic shut off when the arm is traveling at 35 IPS.

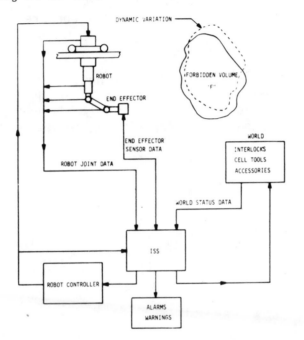

Figure 12 - Intelligent Safety System (ISS)

An intelligent safety system that monitors the robot position, the sensor mounted on the arm and sensors in the surrounding environment is in progress. The systems diagram is shown in Figure 12.

Further information on the use of robotics by NASA can be found in a paper by Fernandez, Jones III and Roberts (4).

CONCLUSIONS

Of the two HT3's one was installed to spray foam insulation, and the other will be placed on the production floor to spray ablator (SLA) by the end of 1984.

In addition, two thermwood gantry manipulators were installed for SLA machining, and a dual head driller/riveter with a 20 feet vertical travel will be used for the intertank assembly.

The major contraints to apply robots for the ET manufacturing are the low volume of production, the large size of the subassemblies and the expensive hardware with which the robots would interact.

BIBLIOGRAPHY

1. Paul M. Lynch, "Economic Technological Modeling and Design Criteria for Programmable Assembly Machines," Doctoral Dissertation, M.I.T. cs. Draper Laboratory Report I-625, June 1976

2. Systems Definition Handbook. Space Shuttle External Tank - Volume 1 NAS 8-303000 Revision A, April 1983

3. Stanley M. Shinners, "Modern Control Systems Theory and Application," Addison-Wesley Publishing Co., 1978

4. K. Fernandez, cs. Jones III, M. L. Roberts. "NASA's use of Robotics in Building the Space Shuttle." 13th ISRI/Robot 7 Conference Proceedings, 1983.

ACKNOWLEDGMENT

The author acknowledges the technical contribution of Donald Frazier, mechanical engineer; Louis Delhom, electronics engineer; Paul Sayer, software engineer; Hugh Marberry, manufacturing engineer; Mike Garrod and Mike Moreau, robot programmers; and many others that contributed to the publication of this paper.

Reprinted from *Research & Development*, August 1984

Testing Begins for Space Station

A MAJOR IMPACT of the Space Shuttle Program will be more direct insight into the long-term effects of space on materials. Some information was obtained when the shuttle crew revitalized the Solar Max satellite, and still more will be gained from the experiments on the Long Duration Exposure Facility (LDEF), placed in orbit on the same flight and scheduled to be recovered during another shuttle mission in early 1985.

Among the test packages on NASA's 30-ft-long (9-m) unmanned spacecraft is one from TRW Inc., Redondo Beach, CA, called the High Voltage Drain Experiment (HVDE). This package will study the effects of constant bombardment of charged particles in Earth's environment on solar cells and thermal insulation systems.

Those represent two of a number of materials being exposed to low-Earth-orbital conditions. (LDEF is orbiting at about 300 mi altitude with a fixed attitude. One end is oriented toward Earth and one toward space, one side facing the direction of orbital travel and one side trailing).

TRW space plasma scientist Dr. William Taylor, principal investigator of HVDE, told *R&D*, "There are a number of materials, mostly coatings, on board, but ours is the only one put at high voltage and high potentials."

There are four objectives of HVDE. They are: to place large numbers of dielectric samples under electric stress in space; to determine their in-space current drainage behavior; to recover, inspect, and further test these samples on Earth; and to specify allowable electric stress levels for the materials as applied to solar array and thermal control coatings for prolonged exposure to space.

The high-voltage samples on board LDEF are divided among two trays, one situated on the leading edge of the spacecraft and the other on the trailing edge. This is done to permit comparing voltage drained in space from plasmas of varying densities.

Since LDEF doesn't have a central power supply, HVDE has its own batteries and power processors to apply up to 1,000 V of electric stress to some of the samples. During the test, the drainage current behavior of the thin dielectric films is determined by having the forward (exposed) face of the film in contact with the space plasma while a bias voltage is applied to the conducting layer on the rear, unexposed, face.

As a control on the experiment, the TRW study, besides having electrically stressed specimens, also provides for "spectator" samples. These are used to determine the effects on samples from merely being present on LDEF.

The two types of materials being tested in the HVDE project, Taylor said, are complete solar cells with their protective surface layer and "Kapton" multilayer plastic film. "We're testing conventional crystalline silicon solar cells with the cover glass material normally used in TRW-produced solar arrays. Kapton is one of the principal components of thermal blankets used on satellites for thermal control."

The main reason for HVDE, he said, is NASA's desire to operate solar array and thermal control systems under higher voltage conditions for improved performance. In solar array studies, Taylor noted, NASA is considering future orbital missions where arrays would operate at levels of up to several hundred volts, considerably higher than at present.

"It's desirable to do that for higher efficiency," Taylor said. "But since arrays never have been operated at those levels, there's fear that effects caused by the space environment might result in efficiency losses at high voltage conditions. Our data might show thresholds for such degradation, in which case we could tell designers to avoid putting arrays at voltage levels above a certain critical value," he said.

High-voltage problems with a thermal film material like Kapton could arise from effects caused by naturally occurring particles, Taylor pointed out. "For instance, in geosynchronous orbit, some of the particles in the Van Allen belt might charge the surface to a very high voltage, posing problems from sparks.

"Such sparks could cause logic upsets—change of states in transistors or switches. If we found this was a problem, there are various ways it might be solved. One approach might be to put a conductive layer around the outside rather than a nonconductive one like 'Teflon'," he explained.

Conversely, there might be some performance advantage that might be gained from using a bias control technique to increase the voltage of the control material. Data obtained from HVDE might indicate the proper bias parameters to use to avoid any difficulties under space conditions.

Of course, it's hoped no unexpected or unsolvable problems are discovered for Kapton at high voltages. Taylor said, "It's possible the data might suggest not using the material. So far, evidence for it at low voltages indicates it's a very stable polymer system. What HVDE is probing is whether it has any stability problems at high voltages."

Taylor and other scientists with experiments on LDEF are eagerly looking forward to beginning Earth-based tests next year. "Not much has been brought back from space yet that has been exposed for long periods of time. In the past, satellites coming back into the atmosphere burned up under the severe reentry conditions," he said.

—*Irwin Stambler* □

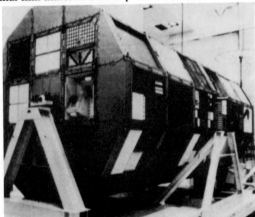

To prepare for the day when the U.S. has a permanent orbiting space station, NASA is testing materials candidates on the Long Duration Exposure Facility. With LDEF, NASA aims to gain insight into the long-term effects of space on materials used for the various structural and electrical components of a space station.—TRW photograph

INDEX

X

Z